PRACTICAL LOW POWER DIGITAL VLSI DESIGN

PRACTICAL LOW POWER DIGITAL VLSI DESIGN

by

Gary Yeap
Motorola

KLUWER ACADEMIC PUBLISHERS
Boston / Dordrecht / London

Distributors for North America:
Kluwer Academic Publishers
101 Philip Drive
Assinippi Park
Norwell, Massachusetts 02061 USA

Distributors for all other countries:
Kluwer Academic Publishers Group
Distribution Centre
Post Office Box 322
3300 AH Dordrecht, THE NETHERLANDS

Library of Congress Cataloging-in-Publication Data

A C.I.P. Catalogue record for this book is available
from the Library of Congress.

Printed on acid-free paper.

Printed in the United States of America

Preface

Low power VLSI design has been a subject of interest in recent years as evident from the surge of related activities in the engineering and research community. The topic has been very much left isolated since the invention of integrated circuits. For long, it was only practiced by a few elite designers in specific areas such as medical electronic implant. It is not until the proliferation of portable consumer electronics that the low power challenges emerge in the mass market digital design. Besides the business demand, the evolution of semiconductor process technology has led to the steady increase in per chip power consumption. For someone who was once an electronic hobbyist like me, the declaration that some commercial microprocessor power dissipation is more than that of a hand held soldering iron certainly wakens the memories of blistered finger burnt. With the Moore's Law showing no signs of deceleration, the power consumption of a VLSI chip is expected to rise beyond the current status. Last but not least, the environmental concerns about the energy consumption of office computing equipment have also fueled the low power needs on the techno-political front.

This book was written for VLSI design engineers and students who have had a fundamental knowledge in CMOS digital design. Part of the contents of this book was conceived when I developed a company wide training class "Tutorial on Low Power Digital VLSI Design" for designers in Motorola. The feedback from the tutorial attendees helps to establish the themes and tone of this book. This book covers the practical low power design techniques and their analysis method at various levels of design abstraction. There is no universal technique that can solve the power dissipation problem in all application areas. It is also my observation that power efficiency cannot be achieved without affecting the other figures of merits of the design. This book emphasizes the optimization and trade-off techniques that involve power dissipation, in the hope that the readers are better prepared the next time they are presented with a low power design problem. It does not document particular results collected from some low power design projects. Rather, it highlights the basic principles, methodologies and techniques that are common to most CMOS digital designs. The advantages and disadvantages of a particular low power technique will be discussed. Besides the classical area-performance trade-off, the impact to design cycle time, complexity, risk, testability, reusability will be discussed. The wide impacts to all aspects of design are what make the low power problems challenging and interesting. Heavy emphasis will be given to top-down structured design style with occasion coverage in the semicustom design methodology. The examples and design techniques cited have been known to be applied to production scale designs or laboratory settings. The goal is to permit the readers to practice the low power techniques using current generation design style and process technology.

Today, the VLSI design task is so huge that specialization is a must. As a result, most designers are experienced in their corresponding focused areas. However, the low power problem is one that calls for total solution at all levels of design abstraction to achieve the highest impact. A design decision made at one level of abstraction can hamper or aid the low power goal in the downstream design process. A breadth of knowledge from all aspects of the design from specification to mass production is required. Hopefully, this book can provide a comprehensive coverage in all areas of the digital design domain. Some analysis techniques have been so mature that commercial Computer-Aided Design software packages have routinely been used in the design process. This book will illuminate the application and the potential role of the software in solving low power problems.

The book is intended to cover wide ranges of design abstraction levels spanning circuit, logic, architecture and system. The art of chip design demands solid intuition, skill and experience. If the craftsmanship of chip design can be acquired through hard work, I believe the basic drill is in the qualitative and quantitative analysis at the various levels of design abstraction. The first three chapters provide enough basic knowledge to cover the qualitative and quantitative analysis at the different design abstraction levels. It is recommended that Chapter 1 thorough 3 be read, in succession, before the later chapters. Subsequent chapters present the low power techniques at the circuit, logic, architecture and system levels. Chapter 6 includes special techniques that are specific to some key areas of digital chip design. The last chapter provides a glimpse of the low power techniques appearing on the horizon.

I would like to thank Prof. Majid Sarrafzadeh of Northwestern University for his encouragement and support during the writing of this book. I would also like to express my sincere gratitude to my friends and colleagues who helped to review the manuscript of this book: Dr. Chih-Tung Chen, Dr. Jenny Gong, Dr. Weiliang Lin, Alberto Reyes, Hongyu Xie.

Gary K. Yeap, Ph.D.
Gilbert, Arizona

Contents

PRACTICAL LOW POWER DIGITAL VLSI DESIGN

CHAPTER 1   *Introduction*

The art of power analysis and optimization of integrated circuits used to be a narrow speciality in analog circuit design. It is now appearing in the mainstream digital design community affecting all aspects of the design process. The interests in low-power chips and systems are driven by both business and technical needs. The industry for low power consumer electronic products is booming with a rapidly expanding market. At the same time, newer generations of semiconductor processing technologies present more stringent requirements to the power dissipation of digital chips due to increased device density, speed and complexity.

Power is the rate at which energy is delivered or exchanged. An analogy of energy and power is water and its flow rate. In a VLSI chip, electrical energy is converted to heat energy during operations. The rate at which energy is taken from the source and converted into heat is the power dissipation. Heat energy has to be dissipated from the chip to avoid an increase in chip temperature, which can cause temporary or permanent circuit failure.

The VLSI low power design problems can be broadly classified into two: *analysis* and *optimization*. Analysis problems are concerned about the accurate estimation of the power or energy dissipation at different phases of the design process. The purpose is to increase confidence of the design with the assurance that the power consumption specifications are not violated. Evidently, analysis techniques differ in their accuracy and efficiency. The accuracy of analysis depends on the availability of design information. In early design phases, the emphasis is to obtain power dissipation estimates rapidly with very little available information on the design. In these phases, less accurate analysis results are expected and tolerated. As the design proceeds to reveal more lower-level details, a more accurate analysis can be performed. Here, better accuracy is demanded and longer analysis time is allowed.

Analysis techniques also serve as the foundation for design optimization. Optimization is the process of generating the best design, given an optimization goal, without violating design specifications. An automatic design optimization algorithm requires a fast analysis engine to evaluate the merits of the design choices. Manual optimization also demands a reliable analysis tool to provide accurate estimation of power dissipation. A decision to apply a particular low power technique often involves trade-offs from different sources pulling in various directions. Major criteria to be considered are the impact to the circuit delay, which affects the performance and throughput of the chip, and the chip area, which directly translates to manufacturing costs. Other factors of chip design such as design cycle time, testability, quality, reliability, reusability, risk, etc., may all be affected by a particular design decision to achieve the low power requirement. Power efficiency cannot be achieved without yielding to one or more of these factors. The task of a design engineer is to carefully weigh each design choice within the specification constraints and select the best implementation.

This chapter starts with a brief introduction to the driving forces behind the low power needs. Before we set out to analyze or optimize the power dissipation of a VLSI chip, the basic understanding of the fundamental circuit theory of power dissipation is imminent. Section 1.2 to 1.5 is a summary of the basic power dissipation modes of a digital chip. We will emphasize the qualitative effects of the power dissipation common to all digital circuits. We shall assume that the readers are familiar with the basic steps of CMOS digital VLSI design, in particular, circuit and logic design, cell library, logic synthesis and commonly used subsystems such as RAMs, adders, multipliers, etc. A good introduction to CMOS VLSI design is given by Weste and Eshraghian [1.1].

1.1 Needs for Low Power VLSI Chips

Power dissipation of VLSI chips is traditionally a neglected subject. In the past, the device density and operating frequency were low enough that it was not a constraining factor in the chips. As the scale of integration improves, more transistors, faster and smaller than their predecessors, are being packed into a chip. This leads to the steady growth of the operating frequency and processing capacity per chip, resulting in increased power dissipation.

There are various interpretations of the *Moore's Law* that predicts the growth rate of integrated circuits. One estimate places the rate at 2X for every eighteen months. Others claim that the device density increases ten-fold every seven years. Regardless of the exact numbers, everyone agrees that the growth rate is rapid with no signs of slowing down. New generations of processing technology are being developed while present generation devices are at a very safe distance from the fundamental physical limits. A need for low power VLSI chips arises from such evolution forces of integration circuits.

Another factor that fuels the needs for low power chips is the increased market demand for portable consumer electronics powered by batteries. The craving for smaller, lighter and more durable electronic products indirectly translates to low power requirements. Battery life is becoming a product differentiator in many portable electronic markets. Being the heaviest and biggest component in many portable systems, batteries have not experienced the similar rapid density growth compared to electronic circuits. The *specific weight* (stored energy per unit weight) of batteries barely doubles in several years. Besides technological issues, further increase in battery specific weight will soon draw safety concerns because the energy density is approaching that of explosive chemicals. It has been generally concluded that the battery technology alone will not solve the low power problem in the near future.

Ironically, high performance computing system characterized by large power dissipation also drives the low power needs. The power dissipation of high performance microprocessors is now approaching several dozen Watts [1.2], comparable to that of a hand-held soldering iron. Power dissipation has a direct impact on the packaging cost of the chip and the cooling cost of the system. Some personal computer's CPUs require cooling fans directly mounted on the chip carriers due to the high power dissipation. A chip that operates at 3.3V consuming 10W means that the average current is 3A. The transient current could be several times larger than the average current. This creates problems in the design of power supply rails and poses big challenges in the analysis of digital noise immunity.

Another major demand for low power chips and systems comes from environmental concerns. Modern offices are now furnished with office automation equipment that consume large amount of power. A study by *American Council for an Energy-Efficient Economy* estimated that office equipment account for 5% of total US commercial energy usage in 1993 and could rise to 10% by the year 2000 if no actions are taken to prevent the trend [1.3]. Computers are the fastest-growing electricity loads in the commercial sector. Since electricity generation is a major source of air pollution, inefficient energy usage in computing equipment indirectly contributes to environmental pollution. The problem has prompted The US Environmental Protection Agency and The US Department of Energy to promote the *Energy Star* program [1.4] that sets guidelines for the energy usage of computing equipment. A similar program has also been adopted worldwide by many industrialized countries. Today, more than two thousand desktop computers, monitors and printers have been certified as Energy Star compliant.

1.2 Charging and Discharging Capacitance

In the next three sections, we will study the fundamental physical mechanisms of power dissipation in digital CMOS VLSI circuits. Some of the most important equations for the analysis of power dissipation will be derived. It is suggested that the readers have a good grasp of the contents of these sections. As we discuss low power analysis and optimization techniques in later chapters, the material in these sections will be referred.

There are two types of power dissipation in CMOS circuits: *dynamic* and *static*. Dynamic power dissipation is caused by switching activities of the circuits. A higher operating frequency leads to more frequent switching activities in the circuits and results in increased power dissipation. Static power dissipation is related to the *logical states* of the circuits rather than switching activities. In CMOS logic, leakage current is the only source of static power dissipation. However, occasional deviations from the strict CMOS style logic can cause static current to be drawn.

The most significant source of dynamic power dissipation in CMOS circuits is the charging and discharging of capacitance. Sometimes, capacitors are intentionally fabricated to achieve certain non-digital operations such as charge sharing and signal delay. However, most digital CMOS circuits do not require capacitors for their intended operations. The capacitance forms due to parasitic effects of interconnection wires and transistors. Such parasitic capacitance cannot be avoided and it has a significant impact on the power dissipation of the circuits. The estimation and analysis of parasitic capacitance is a crucial subject matter not only for signal delay, but also for power.

Figure 1.1 depicts the equivalent circuit of charging and discharging output capacitance of a CMOS logic gate. We use a switch that flips up and down to model the charging and discharging cycles. Referring to the figure, V is an ideal constant voltage source and R_c (R_d) is the resistance of the charging (discharging) circuitry, either intentional or parasitic. According to the laws of physics, the voltage $v_c(t)$ and the current $i_c(t)$ of a capacitance C_L at time t are given by

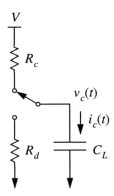

FIGURE 1.1 Equivalent circuit for charging and discharging a capacitor.

$$i_c(t) = C_L \frac{dv_c(t)}{dt} \qquad (1.1)$$

During the charging cycle from time t_0 to t_1, the energy E_s drawn from the voltage source is

$$E_s = \int_{t_0}^{t_1} V i_c(t)\, dt \qquad (1.2)$$

Initially the capacitor contains no charge and the voltage across its terminals is zero, i.e., $v_c(t_0) = 0$. Assume that the capacitor is *fully charged* at the end of the charging cycle, we have $v_c(t_1) = V$. Substituting Equation (1.1) into (1.2), we have

$$E_s = C_L V \int_{t_0}^{t_1} \frac{dv_c(t)}{dt}\, dt = C_L V \int_0^V dv_c = C_L V^2 \qquad (1.3)$$

Part of the electrical energy E_s drawn from the voltage source is stored in the capacitor and the rest is dissipated as heat energy in the resistor R_c. The energy E_{cap} stored in the capacitor at the end of the charging cycle is

$$E_{cap} = \int_{t_0}^{t_1} v_c(t)\, i_c(t)\, dt = C_L \int_{t_0}^{t_1} v_c(t) \frac{dv_c(t)}{dt}\, dt$$

$$= C_L \int_0^V v_c\, dv_c = \frac{1}{2} C_L V^2 \qquad (1.4)$$

From Equations (1.3) and (1.4), the energy E_c dissipated at R_c during charging is therefore

$$E_c = E_s - E_{cap} = \frac{1}{2}C_L V^2 \qquad (1.5)$$

Now consider the discharging cycle from t_1 to t_2, we assume that the capacitor is *fully discharged*, i.e., $v_c(t_1) = V$ and $v_c(t_2) = 0$. The energy E_d dissipated in the discharge resistor R_d is

$$E_d = -\int_{t_1}^{t_2} v_c(t) i_c(t)\, dt = -C_L \int_V^0 v_c(t)\, dv_c = \frac{1}{2}C_L V^2 \qquad (1.6)$$

E_d is exactly equal to the energy stored in the capacitance at the beginning of the discharging cycle. If we charge and discharge the capacitance at the frequency of f cycles per seconds, the power dissipation of the system is

$$P = E_s f = C_L V^2 f \qquad (1.7)$$

Equation (1.7) is the most important power dissipation equation in digital VLSI chip design. It relates power dissipation to a few quantities that are readily observable and measurable in VLSI circuits. The equation is easy to compute, and more important, it is applicable to almost every digital CMOS circuit because only a few mild assumptions have been used in the derivation. This should be evident in the later chapters when we apply the golden equation numerous times throughout this book.

We now discuss the generality and applicability of the $P = C_L V^2 f$ equation by reviewing the derivation procedure. We observe that during charging, $C_L V^2$ energy is drawn from the energy source, half of which is dissipated in the charging resistance R_c and the other half is stored in the capacitor. During discharge, the energy stored in the capacitor is dissipated as heat in the discharging resistor R_d. Notice that only a few assumptions are made in our derivation:

1. The capacitance C_L is constant.
2. The voltage V is constant.
3. The capacitor is fully charged and discharged, i.e., $v_c(t_0) = 0$, $v_c(t_1) = V$, $v_c(t_2) = 0$.

The result is independent of the charging and discharging circuitry R_c, R_d; the length of charging and discharging cycle t_0, t_1, t_2; the voltage or current waveform $v_c(t)$, $i_c(t)$, etc. Furthermore, R_c, R_d can be nonlinear, time varying resistance (such as a transistor) and the above derivation is still valid as long as the three assumptions are satisfied. Most CMOS digital circuits that we design today satisfy the three assumptions above when their capacitances are charged and discharged during logic operations. For most CMOS circuits operating at medium to high frequency, this is the prevailing mode of power dissipation.

Equation (1.7) is only the power dissipation caused by a single capacitor C_L. In general, the total power should be summed over each capacitance C_i in a circuit yielding

$$P = \sum_i C_i V_i^2 f_i \qquad (1.8)$$

where V_i is the voltage swing across the capacitor C_i switching at frequency f_i. For CMOS circuits, V is typically the same for all capacitance C_i. One simple approximation is to assume that f_i is constant, for example, by taking the average of all f_i's. This allows us to write

$$P = V^2 f \sum_i C_i = C_{total} V^2 f \qquad (1.9)$$

in which C_{total} is the sum of all capacitance, f is the average frequency and V is the voltage swing.

In today's typical CMOS process with minimum feature size of 0.18-0.5um, typical values of C_i are in the order of 0.005pF to 1pF. The charging and discharging frequency can be as high as several hundred MHz, with V at several volts.

Example:

A 32 bit off-chip bus operating at 5V and 66MHz clock rate is driving a capacitance of 25pF/bit. Each bit is estimated to have a toggling probability of 0.25 at each clock cycle. What is the power dissipation in operating the bus?

$$C = 32 \times 25 = 800\text{pF}$$
$$V = 5.0\text{V}$$
$$f = 0.25 \times 66 = 16.5\text{MHz}$$
$$P = 800\text{pF} \times 5^2 \text{V}^2 \times 16.5\text{MHz} = 330\text{mW}$$

$\square$

Example:

The chip size of a CPU is 15mm $\times$ 25mm with clock frequency of 300MHz operating at 3.3V. The length of the clock routing is estimated to be twice the circumference of the chip. Assume that the clock signal is routed on a metal layer with width of 1.2um and the parasitic capacitance of the metal layer is $1\,fF/um^2$. What is the power dissipation of the clock signal?

$$C = 4\,(15+25)\,mm \times 1.2um \times 1\,fF/um^2 = 192pF$$
$$V = 3.3\,V$$
$$f = 300MHz$$
$$P = CV^2f = 192pF \times 3.3\,V^2 \times 300MHz = 627mW$$

$\square$

1.3 Short-circuit Current in CMOS Circuit

In the previous section, we have seen the primary mode of power dissipation in a CMOS circuit, i.e., charging and discharging of capacitance. This component of power dissipation is dominant for high frequency CMOS circuits. In this section, we will examine another component of power dissipation also caused by signal switching called *short-circuit power*.

1.3.1 Short-circuit Current of an Inverter

Figure 1.2 shows a simple CMOS inverter operating at V_{dd} with the transistor threshold voltages of V_{tn} and V_{tp} as marked on the transfer curve. When the input signal level is above V_{tn}, the N-transistor is turned on. Similarly, when the signal level is below V_{tp}, the P-transistor is turned on. When the input signal v_i switches, there is a short duration in which the input level is between V_{tn} and V_{tp} and both transistors are turned on. This causes a short-circuit current from V_{dd} to ground and dissipates power. The electrical energy drawn from the source is dissipated as heat in the P and N-transistors.

From the first order analysis of the MOS transistor model, the time variation of the short-circuit current during signal transition is shown in Figure 1.3. The current is zero when the input signal is below V_{tn} or above V_{tp}. The current increases as v_i rises beyond V_{tn} and decreases as it approaches V_{tp}. Since the supply voltage is constant, the integration of the current over time multiplies by the supply voltage is the energy dissipated during the input transition period.

The shape of the short-circuit current curve is dependent on several factors:

1. The duration and slope of the input signal.

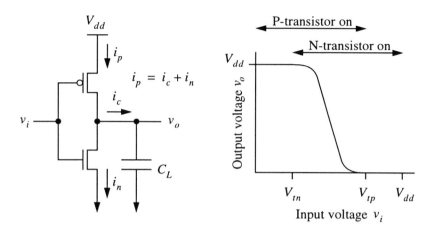

FIGURE 1.2 A CMOS inverter and its transfer curve.

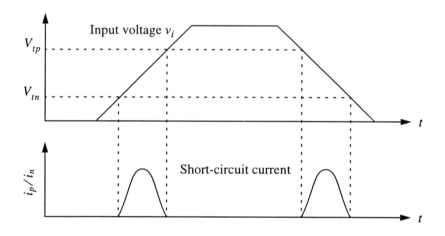

FIGURE 1.3 Short-circuit current of a CMOS inverter during input transition.

2. The I-V curves of the P and N transistors, which depend on their sizes, process technology, temperature, etc.

3. The output loading capacitance of the inverter.

Under an input ramp signal, a first order analysis of the short-circuit current [1.1] reveals that the energy dissipated is

$$E_{short} = \frac{\beta}{12}\tau(V_{tp} - V_{tn})^3 \qquad (1.10)$$

where β is the size of the transistors and τ is the duration of the input signal. This equation is useful in showing the relationship of each parameter to the short-circuit energy or power dissipation but is seldom used to compute the actual value. In real circuits, the actual energy dissipation is a complex function of the factors listed above. Furthermore, the equation assumes that the output loading capacitance of the CMOS inverter is zero which is not true in real circuits.

Despite the difficulties in expressing the short-circuit current and energy, they can be computed quite accurately using circuit simulators such as SPICE. Many commercially available circuit simulation tools can simulate the short-circuit power dissipation accurately. To obtain the short-circuit current when the inverter input changes from low to high, we simply measure the source-drain current of the N-transistor. If the input changes from high to low, we measure current through the P-transistor. As a rule of thumb, this mode of power dissipation is generally 10-60% of the total power dissipation of a CMOS gate. In a cell-based design in which all basic circuit elements are pre-designed, the short-circuit power dissipation can be characterized much like timing delay, as will be discussed in Section 2.3.3. For a full-custom design style in which cells are not characterized, transistor-level simulation can be used to measure the power dissipation with reasonable accuracy and speed; see Section 2.1.2.

1.3.2 Short-circuit Current Variation with Output Load

Short-circuit current exhibits some interesting characteristic with respect to the output loading capacitance and the input signal slope of a CMOS inverter. These characteristics have some implications on the analysis and optimization of the power efficiency for CMOS circuits. We have observed that duration of short-circuit current depends on the transition period of the input signal. The short-circuit current has also been shown to depend on the output loading capacitance [1.5].

The qualitative effects of the output loading on short-circuit current can be analyzed in a thought experiment. Consider the case when the input voltage is falling and the output voltage is rising. Based on the on-off properties of the transistors, the short-circuit current is non-zero only when the input level is between V_{tn} and V_{tp}. When the inverter input is at $0.5V_{dd}$, its output voltage is somewhere between zero and $0.5V_{dd}$ assuming a symmetrical transfer curve. If the output capacitance is large, the output voltage barely rises above zero and the voltage across the source and drain of the N-transistor is only slightly above zero. The low source-drain potential difference results in small short-circuit current. Conversely, if the output capacitance is small, the output voltage rises faster and the source-drain voltage is much higher, causing a larger short-circuit current.

To see the effects of output capacitance on short-circuit current, we ran several SPICE simulations with different capacitance values while keeping all conditions (signal slope, temperature, etc.) constant. The short-circuit currents are plotted in Figure 1.4.[*]

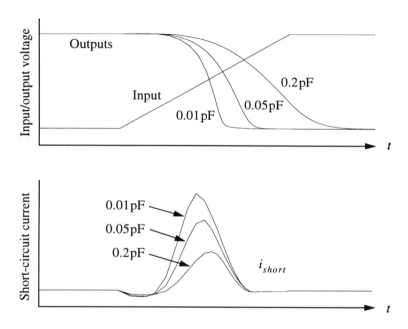

FIGURE 1.4 Short-circuit current under varying output capacitance.

We observe that the short-circuit current envelope is the largest when the output capacitance is the smallest. As the output capacitance increases, the current envelope becomes smaller and eventually approaches zero. However, the *duration* of the short-circuit current is independent of the output capacitance because it depends on the input signal slope only. The short-circuit current is non-zero only when the input voltage is between V_{tn} and V_{tp}. Thus if all conditions are kept identical, increasing the output loading capacitance has the effect of reducing the short-circuit energy per transition.

---

* Figures 1.4 - 1.6 are generated with MCSPICE, a Motorola proprietary circuit simulator.

From the above observation, one might question whether increasing the output loading capacitance can reduce power dissipation of the circuit because of smaller short-circuit current; or is there an optimum output loading value that optimizes the total power dissipation of a CMOS gate? The answer is negative as shown by Figure 1.5. In this experiment, we again keep all conditions identical and vary only the output capacitance. This time, we measure the total current, i.e., the sum of short-circuit current and the capacitance charging or discharging current.

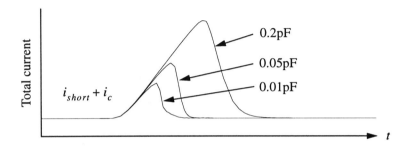

FIGURE 1.5 Total current under varying output capacitance.

The results show that when the output capacitance is increased, the total current always increases in peak as well as duration. This means that increasing the output capacitance decreases the short-circuit power dissipation but the sum of capacitive and short-circuit power increases. Therefore, capacitance is always the enemy of power efficiency and a low power digital design should always strive to reduce loading capacitance. A major contribution of capacitance is the input capacitance of the signal receiving gate. The size of the receiving gate is constrained by its speed requirement. Driving a signal faster than necessary results in wasted power. Thus for low power design, a good design practice is to choose the minimum gate size that meets the speed constraint to minimize capacitance. This agrees with the minimum area design goal and presents no conflicting requirements in the area-power trade-off.

To understand the effects of increasing output capacitance on short-circuit current (i_{short}), capacitor current (i_c) and total current ($i_{short} + i_c$), we refer to the current envelopes of the SPICE simulations shown in Figure 1.4 and Figure 1.5. We note the *width* (the duration in which the current is non-zero), *peak* and the *integration* of the current envelope over time, which corresponds to the energy dissipated per transition. The observations are summarized in the following table:

TABLE 1.1 **Effects of increasing output loading capacitance.**

Current envelope	Width	Peak	Integration
i_{short}	unchanged	decrease	decrease
i_c	increase	increase	increase
$i_{short} + i_c$	increase	increase	increase

1.3.3 Short-circuit Current Variation with Input Signal Slope

A more interesting analysis of short-circuit current is its variation with the input signal slope. Again, we perform several SPICE circuit simulations in which all conditions are kept identical except the input signal slope. The short-circuit current and the energy dissipation is plotted in Figure 1.6.

The results clearly show that as the input signal slope deteriorates (longer signal transition time), the short-circuit current peak and duration increase. This is also confirmed by Equation (1.10). For the total current, i.e., short-circuit plus capacitance current, the width increases but the peak decreases. However, the integration of the total current over time (energy) is increasing. The capacitive charging/discharging energy remains identical because the capacitance is kept constant in all experiments. The observations are summarized in Table 1.2 below.

TABLE 1.2 **Effects of deteriorating input signal slope.**

Current envelope	Width	Peak	Integration
i_{short}	increase	increase	increase
i_c	increase	decrease	unchanged
$i_{short} + i_c$	increase	decrease	increase

The results tell us that to improve power efficiency of a CMOS circuit, we should use the fastest input signal slope for all signals. This immediately poses a complicated optimization issue: while the sharp signal slope reduces power dissipation of the signal *receiving* gate, the transistor sizes of the signal *driving* gate have to be increased to sustain the steep slope. A larger driver gate means more power consumption due to increased input capacitance and short-circuit current. Thus, enforcing sharp signal slopes presents conflicting effects on the power dissipation of the signal driver and

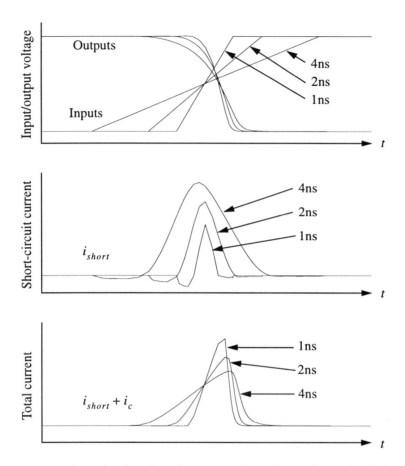

FIGURE 1.6 Short-circuit and total current under different input signal slopes.

receiver gates. Furthermore, the issue is complicated by the delay requirements of the gates to satisfy performance constraints. There is no simple answer to this problem and most chip designers eventually retreat to some rules of thumb. The general rule is that the duration of the input signal (e.g., as measured by 10% to 90% voltage threshold) should be comparable to the propagation delay of the gate. For example, if the typical gate delay is approximately 0.75ns, a signal slope constraint of 0.5ns to 1.0ns is appropriate. For special cases such as off-chip signals or bus drivers, some adjustment to the rule may be necessary to balance the delay, area and power requirements.

1.4 CMOS Leakage Current

Leakage current exists as a natural phenomenon of the semiconductor device operation. Leakage is a form of current that is generally not intended for the normal operation of a digital circuit. Generally, leakage current serves no useful purposes although some circuits do exploit it for intended operations, such as power-on reset signal generation. In circuits that use MOS transistors, there are two major sources of leakage current: 1. reverse biased PN-junction current and, 2. subthreshold channel conduction current.

1.4.1 Reverse Biased PN-junction

The first source of leakage current occurs when the source or drain of an N-transistor (P-transistor) is at V_{dd} (*Gnd*). PN-junctions are formed at the source or drain of transistors because of a parasitic effect of the bulk CMOS device structure. As shown in Figure 1.7, the junction current at the source or drain of the transistor is picked up through the bulk or well contact. The magnitude of the current depends on the temperature, process, bias voltage and the area of the PN-junction. Analysis of the semiconductor physics shows that the reverse biased PN-junction current is

$$I_{reverse} = I_s (e^{V/V_{th}} - 1) \tag{1.11}$$

The current I_s is the reverse saturation current dependent on the fabrication process and the PN-junction area. The variable V_{th}, which often appears in leakage equations, is called the *thermal voltage* where

$$V_{th} = kT/q \tag{1.12}$$

In the above equation, $k = 1.38 \times 10^{-23}$ Joule/K is the Boltzmann's constant, $q = 1.60 \times 10^{-19}$C is the electronic charge and T is the device operating temperature. At room temperature, $T = 300$K and $V_{th} = 25.9$mV.

Note that Equation (1.11) is applied when V is negative, i.e., the PN-junction is in reversed bias. If $|V| \gg V_{th}$, we have $e^{V/V_{th}} \approx 0$ and $I_{reverse} = I_s$. Thus, a small reverse voltage is sufficient to induce current saturation. For all practical purposes, the current is largely independent of the circuit operating voltage.

The reverse saturation current I_s is of the order of 1pA/um^2 for today's process. Thus, for a two-million-transistor chip with an average junction area of 1×10um^2 per transistor, the current is about 20uA. This current is too small compared to the dynamic current. For example, an inverter driving a capacitor of 0.5pF operating at 3.3V, 20MHz will consume $I = CVf = 0.5$pF $\times 3.3$V $\times 20$MHz $= 33$uA, which is

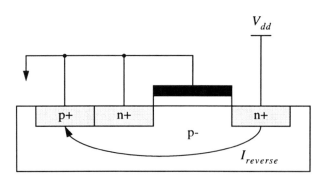

FIGURE 1.7 Reverse biased PN-junction current.

larger than the junction leakage of all two million transistors. The saturation current I_s is highly dependent on the temperature. As a rule of thumb, the saturation current I_s doubles for every ten-degree increase in temperature.

To summarize, the reverse biased PN-junction current is largely independent of operating voltage but depends on the fabrication process, junction area and temperature. Generally, the leakage current is not a problem for most chips and there is very little one can do to reduce this current in a large scale digital design.

1.4.2 Subthreshold Channel Leakage

The second source of leakage current is the subthreshold leakage through a MOS device channel. Even though a transistor is logically turned off, there is a non-zero leakage current through the channel at the microscopic level, as illustrated in Figure 1.8. This current is known as the *subthreshold leakage* because it occurs when the gate voltage is below its threshold voltage.

Other than the device dimension and fabrication process, the magnitude of the sub-threshold conduction current I_{sub} further depends on: gate voltage V_{gs}, drain voltage V_{ds} and temperature. The dependency on V_{ds} is insignificant when it is much larger the thermal voltage. In CMOS circuits, $V_{ds} \approx V_{dd}$ when subthreshold current occurs and thus, the current essentially depends on V_{gs}, device dimension and operating temperature.

Subthreshold conduction current is given by [1.6]

$$I_{sub} = I_0 \, e^{(V_{gs} - V_t)/(\alpha V_{th})} \tag{1.13}$$

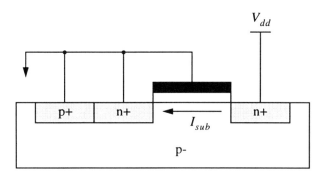

FIGURE 1.8 Subthreshold leakage current of a MOS transistor.

where V_t is the device threshold voltage; V_{th} is the thermal voltage (V_{th} = 25.9mV at room temperature) as defined in Equation (1.12); and I_0 is the current when $V_{gs} = V_t$. The parameter α is a constant depending on the device fabrication process, ranging from 1.0 to 2.5. The exponent factor ($V_{gs} - V_t$) has a negative value so that I_{sub} drops exponentially as V_{gs} decreases.

A typical plot of I_{sub} with V_{gs} of a MOS transistor is illustrated in Figure 1.9. The slope at which I_{sub} decreases is an important parameter in low power design. By taking the logarithm and differentiation of Equation (1.13), the slope is conveniently expressed as

$$\frac{\Delta V_{gs}}{\Delta(\log I_{sub})} = \frac{\alpha V_{th}}{\log e} = 2.3\alpha\frac{kT}{q} \tag{1.14}$$

At room temperature when V_{th} = 25.9mV, the slope ranges from an ideal 60mV/decade to 150mV/decade. A smaller value is desirable because the subthreshold current decreases more per unit changes in gate voltage V_{gs}. The slope flattens as the temperature T rises, as indicated by Equation (1.14).

Subthreshold conduction current is becoming a limiting factor in low voltage and low power chip design. When the operating voltage is reduced, the device threshold voltage V_t has to be reduced accordingly to compensate for the loss in switching speed. Consider a conventional CMOS circuit operating at V_{dd} = 5.0V and V_t = 0.9V. Assuming an ideal slope of 60mV/decade in Equation (1.14), the subthreshold current is about 900/60 = 15 decades smaller from $V_{gs} = V_t$ to V_{gs} = 0. Such low leakage generally does not arouse design concern. However, in an aggressive low power design that requires low operating voltage, the voltages might be V_{dd} = 1.2V, V_t = 0.45V. This translates to only 450/60 = 7.5 decades difference, 7.5 orders of

17

magnitude increase in subthreshold conduction current as compared to the high voltage device. The problem is worsened by the temperature factor in Equation (1.14). At high temperature, e.g., 125 degree Celsius, the slope becomes 80mV/decade and the subthreshold conduction current could deteriorate to $450/80 = 5.6$ decades.

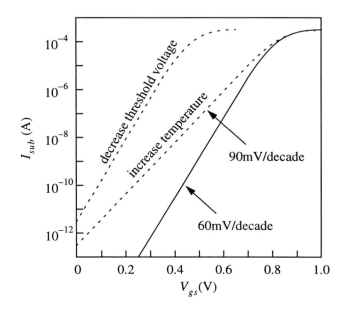

FIGURE 1.9 Subthreshold current as a function of gate voltage.

To illustrate this graphically in Figure 1.9, decreasing the device threshold voltage V_t shifts the ideal 60mV/decade I_{sub} curve to the left while increasing temperature flattens the slope, both lead to increased subthreshold conduction current [1.7]. Another disappointing factor is the variation of V_t due to the device fabrication process. When the threshold voltage is scaled down, the variation does not scale accordingly. As a result, the worst case subthreshold leakage current increases dramatically for low threshold devices. This trend could reach a point where it becomes a limiting factor at very low operating voltages (below 1.0V). Methods to reduce subthreshold leakage by transistor sizing and threshold voltage adjustment have been reported (see Section 4.6).

1.4.3 Leakage Current in Digital Design

Subthreshold leakage and reverse-biased junction leakage have very similar characteristic. They are both in the order of pico-Ampere per device and very sensitive to process variation. Both increase dramatically with temperature and are relatively independent of operating voltage for a given fabrication process. Although the leakage current cannot be ignored in some cases, the logic or circuit designers typically do not pay any special attention to attempt to reduce the current. The reason is that the leakage current is generally beyond the digital designer's control. Leakage current is difficult to predict, measure or optimized. Most large scale, high performance digital chips operate in high frequency regions in which the dynamic power dissipation is several million times larger than the leakage current. For small custom circuits, especially analog circuits, in which the designers have full control over the layout and transistor sizes, leakage current analysis can be performed using SPICE simulation or simple rule of thumb calculation. However, as we move toward next generation CMOS fabrication processes with very low device threshold and operation voltage, leakage current may become a limiting factor. This poses low power challenges for device and process designers.

1.5 Static Current

Strictly speaking, digital CMOS circuits are not supposed to consume static power from constant static current flow. All non-leakage current in CMOS circuits should only occur in transient when signals are switching. However, there are times when deviations from CMOS style circuit design are necessary.

An example is the pseudo NMOS logic circuit, as shown in Figure 1.10. In general, the pseudo NMOS circuit is not used on random logic. However, for special circuits such as PLAs or Register Files, it may be useful due to its efficient area usage. In such a circuit, there is a constant current flow from V_{dd} to *Gnd* which dissipates power. This is an example where power is traded for area efficiency.

The pseudo NMOS circuit does not require a P-transistor network and saves half the transistors required for logic computation as compared to the CMOS logic. The circuit has a special property that the current only flows when the output is at logic 0. When the output is at logic 1, all N-transistors are turned off and no static power is consumed, except leakage current. This property may be exploited in a low power design. If a signal is known to have very high probability of logic 1, say 0.99, it may make sense to implement the computation in pseudo NMOS logic. Conversely, if the

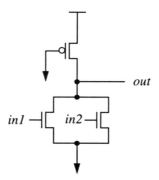

FIGURE 1.10 A pseudo NMOS NOR gate.

signal probability is very close to zero, we may eliminate the N-transistor network of a CMOS gate and replace it with a load transistor of N type.

An example where this feature can be exploited is the system reset circuitry. The reset signal has extremely low activation probability (for example, during the power-on phase) which can benefit from such circuit technique. Other examples where signal activation probabilities are extremely low are: test signals, error detection signals, interrupt signals and exception handling signals.

1.6 Basic Principles of Low Power Design

The previous sections have provided the basic understanding of the power dissipation of digital VLSI chips. Equations (1.8), (1.10), (1.11) and (1.13) express the fundamental modes of power dissipations in CMOS VLSI circuits. For any low power technique encountered, we can always trace its effect to the above equations. By examining the equations, several basic principles of low power design techniques can be uncovered. In this section, we will describe some common themes expressed in some of the low power techniques found in this book. We hope that the study of the basic principles can help the readers to develop a global picture of the low power solution techniques and inspire new low power ideas.

Conservation and *trade-off* are the philosophy behind most low power techniques. The conservation school attempts to reduce power that is wasted without a due course. The design skills required are in identifying, analyzing and preventing waste. Another methodology of low power techniques is to explore alternate designs that favor power dissipation. This often requires complex trade-off decisions involving a designer's skill. Overall, intimate understanding of the design specification, operating environment and intuition acquired from past design experience are keys to creative low power techniques. Design techniques and methodologies for low power VLSI have been widely reported [1.8] [1.9] [1.10] [1.11] [1.12] [1.13] [1.14].

It should be emphasized that no single low power technique is applicable to all situations. Design constraints should be viewed from all angles within the bounds of the design specification. Low power considerations should be applied at all levels of design abstraction and design activities. Chip area and speed are the major trade-off considerations but a low power design decision also affects other aspects such as reliability, design cycle time, reusability, testability and design complexity. Early design decisions have higher impact to the final results and therefore, power analysis should be initiated early in the design cycle. Maintaining a global view of the power consumption is important so that a chosen technique does not impose restrictions on other parts of the system to offset its benefits.

1.6.1 Reduce Switching Voltage

The dynamic power of digital chips expressed by Equation (1.8) is generally the largest portion of power dissipation. The $P = CV^2f$ equation consists of three terms: voltage, capacitance and frequency. Due to the quadratic effect of the voltage term, reducing the switching voltage can achieve dramatic savings. The easiest method to achieve this is to reduce the operating voltage of the CMOS circuit. Other methods seek to reduce voltage swing by using well-known circuit techniques such as charge sharing, transistor threshold voltage, etc.

There are many trade-offs to be considered in voltage reduction. Performance is lost because MOS transistors become slower at lower operating voltages. The main reason is that the threshold voltages of the transistors do not scale accordingly with the operating voltage to avoid excessive leakage current. Noise immunity is also a concern at low voltage swing. Special level converters are required to interface low-swing signals to the full-swing ones.

1.6.2 Reduce Capacitance

Reducing parasitic capacitance in digital design has always been a good way to improve performance as well as power. However, a blind reduction of capacitance may not achieve the desired result in power dissipation. The real goal is to reduce the product of capacitance and its switching frequency. Signals with high switching frequency should be routed with minimum parasitic capacitance to conserve power. Conversely, nodes with large parasitic capacitance should not be allowed to switch at high frequency. Capacitance reduction can be achieved at most design abstraction levels: material, process technology, physical design (floorplanning, placement and routing), circuit techniques, transistor sizing, logic restructuring, architecture transformation and alternative computation algorithms.

1.6.3 Reduce Switching Frequency

For the sake of power dissipation, the techniques for reducing switching frequency have the same effect as reducing capacitance. Again, frequency reduction is best applied to signals with large capacitance. The techniques are often applied to logic level design and above. Those applied at a higher abstraction level generally have greater impact. Reduction of switching frequency also has the side effect of improving the reliability of a chip as some failure mechanism is related to the switching frequency. One effective method of reducing switching frequency is to eliminate logic switching that is not necessary for computation. Other methods involve alternate logic implementation since there are many ways to design a logic network to perform an identical function. The use of different coding methods, number representation systems, counting sequences and data representations can directly alter the switching frequency of a design.

1.6.4 Reduce Leakage and Static Current

Leakage current, whether reverse biased junction or subthreshold current, is generally not very useful in digital design. However, designers often have very little control over the leakage current of the digital circuit. Fortunately, the leakage power dissipation of a CMOS digital circuit is several orders of magnitude smaller than the dynamic power. The leakage power problem mainly appears in very low frequency circuits or ones with "sleep modes" where dynamic activities are suppressed. Most leakage reduction techniques are applied at low-level design abstraction such as process, device and circuit design. Memory chips that have very high device density are most susceptible to high leakage power.

Static current can be reduced by transistor sizing, layout techniques and careful circuit design. Circuit modules that consume static current should be turned off if not used. Sometimes, static current depends on the logic state of its output and we can consider reversing the signal polarity to minimize the probability of static current flow.

1.7 Low Power Figure of Merits

How do we measure the power efficiency of a chip or system? How do we assess the effectiveness of a particular low power technique, logic family, or design style? Such questions of seemingly technical nature often invoke controversy over the choice figure of merits in the measuring of power efficiency.

Many different units of measure have been used to quantify the figure of merits of a low power chip or system. The most straightforward unit of measure is the *power consumption* in *Watt*. Watt is the absolute measure of power consumed by a chip or a system and often used in total power specification. This is useful for packaging considerations, system power supply and cooling requirements. Also the *peak power*, which is the maximum power consumption of a chip at any time, is used for power ground wiring design, signal noise margin and reliability analysis.

A chip that operates at a higher frequency can perform faster computation and therefore should consume more power. Therefore when we need to measure the power efficiency of a chip, Watt is no longer useful. Since power is the rate at which energy is consumed over time, the measure of *energy* in the unit of *Joule* becomes another choice of measure. The commonly quoted uW/MHz numbers for microprocessors are a measure of energy dissipation per clock cycle, perceived as the power efficiency of the chips. A lower energy number is more desirable because it requires less power to perform computation at the same frequency.

When comparing two processor chips with different instruction sets or architecture, the uW/MHz measure may be misleading. Different processors require different number of clock cycles for the same instructions. This prompts many people to use a more objective measure such as uW/MIPS or mA/MIPS (MIPS means million instruction per second). MIPS is a measure of the performance level of a processor. Normalizing the power dissipation with respect to MIPS allows us to compare processors with widely different performance rating. The uW/MIPS measure has a unit of Watt-Second per Instruction. Thus, it is a measure of the energy consumed by a typical instruction. This figure of merits is useful when comparing power efficiency of processors with similar instruction sets, for example, two DSP processors of the same family. Because of the normalization, this measure is independent of the performance or clock rate of the processor.

When comparing processors with different architectures, uW/MIPS suffers from the same problem as the uW/MHz measure. A Reduced Instruction Set Computer (RISC) requires more instructions than a Complex Instruction Set Computer (CSIC) to complete a given computation task. Recent processor architecture such as the Very Long Instruction Word (VLIW) can actually accomplish many concurrent operations with a single instruction. Thus, a RISC machine will show a lower uW/MIPS number while it may very well be an inferior computation engine because it takes more instructions to achieve the same computation task. This leads to other measures by substituting MIPS with better measures of performance rating such as SPEC, resulting in uW/SPEC. SPEC is a measure of computation speed derived from executing some standard benchmark software programs written in machine independent high-level programming language.

Both uW/MIPS and uW/SPEC measure the energy consumption of a typical instruction or operation regardless of the processor's performance rating. So, they tend to favor machines with simple architecture. Unlike simple machines, high performance machines require additional overhead such as pipelining, instruction restart, etc., which increases the energy consumption per instruction. Therefore, some researchers argue that we should take the instruction energy measure and normalize it with the performance rating. This results in the $uW/MIPS^2$ or $uW/SPEC^2$ measures, which may seem quite artificial at first glance. However, such measures merely place more emphasis on the performance rating than power dissipation by squaring the performance measure.

The uW/MIPS and $uW/MIPS^2$ measures encompass the merits of chip fabrication technology as well as architecture and circuit design techniques. If we wish to factor out the fabrication technology and compare only the merits of the processor design, we can again normalize this measure with technology parameters such as the typical gate delay.

Other measures such as the well-known *energy delay product* is commonly used to assess the merits of a logic style. When battery life is of concern in portable electronics applications, we often define measures that involve the battery lifetime. Commercial batteries are most often rated with mA-Hour, which is a unit of stored energy when the operating voltage is constant.

While the above discussion pointed out several popular figures of merits for low power design, it should be noted that many other parameters are used for different purposes. The choice of the figure of merits depends on the type of analysis and application area. No single measure is necessary for a particular situation nor is it sufficient for all purposes. It is the designer's responsibility to define the appropriate parameters for his analysis and optimization problem.

Further Reading

N. Weste and K. Eshraghian, *Principles of CMOS VLSI Design: A System Perspective*, Second Edition, Addison-Wesley, 1993.

A. Chandrakasan and R. Brodersen, *Low Power Digital CMOS Design*, Kluwer Academic Publishers, Boston, 1995.

A. Bellaouar and M. Elmasry, *Low Power Digital VLSI Design - Circuits and Systems*, Kluwer Academic Publishers, Boston, 1995.

J. Rabaey and M. Pedram, *Low Power Design Methodologies*, Kluwer Academic Publishers, Boston, 1996.

A. Chandrakasan and R. Brodersen, *Technologies for Wireless Computing*, Kluwer Academic Publishers, Boston, 1996.

M. Smith, *Application-Specific Integrated Circuits*, Addison-Wesley, New York, 1997.

REFERENCES

[1.1]   N. Weste and K. Eshraghian, *Principles of CMOS VLSI Design: A System Perspective*, Second Edition, Addison-Wesley, 1993.

[1.2]   D. Bursky, "Power Reduction Schemes Promise Cool Digital ICs," *Electronic Design*, vol. 43, no. 1, Jan. 1995.

[1.3]   "Energy Star Computers, Introducing the Energy Star Computers Program," United States Environmental Protection Agency, EPA 430-F93-049, Nov. 1993. Also in *http://es.inel.gov/partners/estar/strcomp1.html*.

[1.4]   "Purchasing An Energy Star Computer," United States Environmental Protection Agency, EPA 430-K-94-006, Jul. 1994. Also in *http://es.inel.gov/partners/estar/strcomp2.html*

[1.5]   H. Veendrick, "Short-Circuit Dissipation of Static CMOS Circuitry and its Impact on the Design of Buffer Circuits," *IEEE Journal of Solid-State Circuits*, vol. 19, no. 8, pp. 468-473, Aug. 1984.

[1.6]   L. Glasser and D. Dobberpuhl, *The Design and Analysis of VLSI Circuits*, Addison-Wesley, 1985.

[1.7]   S. Sze, *VLSI Technology*, Second Edition, McGraw-Hill, New York, 1988.

[1.8]   A. Chandrakasan, S. Sheng and R. Brodersen, "Low-power CMOS Digital Design," *IEEE Journal of Solid-State Circuits*, vol. 27, no. 4, pp. 473-484, Apr. 1992.

[1.9]   D. Singh, J. Rabaey, M. Pedram, F. Catthoor, S. Rajgopal, N. Sehgal and T. Mozdzen, "Power Conscious CAD Tools and Methodologies: A Perspective," *Proceedings of the IEEE*, vol. 83, no. 4, pp. 570-593, Apr. 1995.

[1.10]   A. Chandrakasan and R. Brodersen, "Minimizing Power Consumption in Digital CMOS Circuits," *Proceedings of the IEEE*, vol. 83, no. 4, pp. 498-523, Apr. 1995.

[1.11]   S. Devadas and S. Malik, "A Survey of Optimization Techniques Targeting Low Power VLSI Circuits," *Proceedings of Design Automation Conference*, pp. 242-247, 1995.

[1.12]   M. Pedram, "Power Minimization in IC Design: Principles and Applications," *ACM Transactions on Design Automation of Electronic Systems*, vol. 1, no. 1, pp. 3-56, Jan. 1996.

[1.13]   G. Yeap and A. Wild, "Introduction to Low-Power VLSI Design," *International Journal of High Speed Electronics and Systems*, vol. 7, no. 2, pp. 1-26, Jun. 1996.

[1.14]   A. Wolfe, "Opportunities and Obstacles in Low-Power System-Level CAD," *Proceedings of Design Automation Conference*, pp. 15-20, 1996.

CHAPTER 2

Simulation Power Analysis

Computer simulation has been applied to VLSI design for several decades. Most simulation programs operate on mathematical models which mimic the physical laws and properties of the object under simulation. Today, simulation is used for functional verification, performance, cost, reliability and power analysis. Many simulation languages have been developed specifically for IC's. For example in digital logic simulation, *VHDL* (Very High Speed IC Hardware Description Language) and *Verilog* are two popular languages being used. Special purpose hardware has also been developed to speed up the simulation process.

Simulation-based power estimation and analysis techniques have been developed and applied to the VLSI design process [2.1] [2.2] [2.3] [2.4] [2.5] [2.6]. Simulation software operating at various levels of design abstraction is a key technology in the mainstream VLSI design. The main difference between simulation at different levels of design abstraction is the trade-off between computing resources (memory and CPU time) and accuracy of the results [2.7]. In general, simulation at a lower-level design abstraction offers better accuracy at the expense of increased computer resource. Circuit simulators such as SPICE attain excellent accuracy but cannot be applied to full-chip analysis. Logic simulation generally can handle full-chip analysis but the accuracy is not as good and sometimes the execution speed is too slow. Behavioral-level or functional-level simulation offers rapid analysis but the accuracy is sacrificed. Figure 2.1 summarizes the trade-off between computing resources and analysis accuracy at different levels of design abstraction.

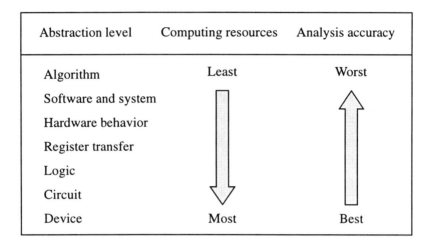

Abstraction level	Computing resources	Analysis accuracy
Algorithm	Least	Worst
Software and system		
Hardware behavior		
Register transfer		
Logic		
Circuit		
Device	Most	Best

FIGURE 2.1 Computing resources and analysis accuracy at various abstraction levels.

Since no single simulation technique is applicable to all levels of design, the top-down estimation, refinement and verification methodology is used. As an example, the designer may start with a simulation at the hardware behavior level to obtain an initial power dissipation estimate. When the gate-level design is available, a gate-level simulation is performed to refine the initial estimate. If the initial estimate turns out to be inaccurate and the design fails the specification, the design is modified and verified again. The iteration continues until the gate-level estimate is within specification. The design is then taken to the transistor or circuit-level analysis to further verify the gate-level estimates. The refinement and verification steps continue until the completion of the design process, when the chip is suitable for mass production.

This chapter is dedicated to simulation techniques to estimate and analyze power dissipation of VLSI chips. The material is presented according to the familiar design abstraction levels. For each technique, we discuss its analysis models, application domains, advantages and disadvantages. The concept of *characterization* will be emphasized. Characterization refers to the process of using lower-level analysis results as a basis to construct higher-level power models. In this manner, the lower-level details can be abstracted to more compact higher-level models to enable more efficient analysis. For example, the SPICE circuit-level analysis can provide data to construct gate-level power models. Subsequently, the gate-level power simulation results are applied to the architecture-level components. The loss of accuracy as we move up to a higher-level abstraction is a natural outcome that cannot be avoided in order to provide faster analysis. It is important for the readers to appreciate the bottom-up propagation of analysis results and the relationship among analyses at different levels of abstraction.

2.1 SPICE Circuit Simulation

SPICE (Simulation Program with IC Emphasis) is the de facto power analysis tool at the circuit level. A wealth of literature has already been dedicated to SPICE and dozens of SPICE-like application software packages are available today. We will only briefly discuss its application to power analysis, assuming that the readers are already familiar with SPICE.

2.1.1 SPICE Basics

SPICE operates by solving a large matrix of nodal current using the Krichoff's Current Law (KCL). The basic components of SPICE are the primitive elements of circuit theory such as resistors, capacitors, inductors, current sources and voltage sources. More complex device models such as diodes and transistors are constructed from the basic components. The device models are subsequently used to construct a circuit for simulation. Basic circuit parameters such as voltage, current, charge, etc. are reported by SPICE with a high degree of precision. Hence the circuit power dissipation can be directly derived from SPICE simulation.

SPICE offers several analysis modes but the most useful mode for digital IC power analysis is called *transient analysis*. The analysis involves solving the DC solution of the circuit at time zero and makes small time increments to simulate the dynamic behavior of the circuit over time. Precise waveforms of the circuit parameters can be plotted over the simulation time.

SPICE device models are derived from a characterization process. Each device model is described by dozens of parameters. The models are typically calibrated with physical measurements taken from actual test chips and can achieve a very high degree of accuracy. Lower-level analysis tools using Finite Element Methods or other physical simulation can also be used to produce the device model parameters.

2.1.2 SPICE Power Analysis

The strongest advantage of SPICE is of cause its accuracy. SPICE is perhaps the most versatile among all power analysis tools. It can be used to estimate dynamic, static and leakage power dissipation. MOS and bipolar transistor models are typically available and it also faithfully captures many low-level phenomena such as charge sharing, cross talk and transistor body effect. In addition, it can handle common circuit components such as diodes, resistors, inductors and capacitors. Specialized circuit components can often be built using the SPICE's modeling capability.

SPICE analysis requires intensive computation resources and is thus not suitable for large circuits. Most SPICE-based simulators start to experience memory or computation limitation at several hundred to several thousand devices. Some advanced SPICE simulators can handle circuits up to ten thousand devices but simulating the entire chip is not possible.

With the correct device models, SPICE simulations can reach an accuracy within a few percent of physical measurement. The main source of error in SPICE is seldom found in the computation process but the inherent difficulties in modeling physical components and devices. Like any mass production process, the fabrication of semiconductor devices is subject to normal fluctuation and therefore SPICE's accuracy is often clouded by the variation of the chip production process. The most common method to cope with this problem is to apply extreme case analysis. Several sets of device models are generated to represent the typical and extreme conditions of the chip fabrication process and operating environment. The conditions are generally defined by the speed of the device so that the designer can predict the fastest and slowest operating extremes of the circuit. For example, most chip designers will simulate SPICE with three sets of parameters, TYPICAL, BEST and WORST case conditions based on the device speed.

The variation of semiconductor process could be large. The BEST and WORST case device speed could be 2X apart. The process variation of power dissipation is less but is still on the same order. For most circuits using conventional CMOS processes, faster device models generally correspond to higher power dissipation and vice versa. However, there are exceptions: for example, a low-speed worst case device model may cause slow signal slopes with high short-circuit power. For designs with marginal power budget, the analysis should be performed on some or all extreme cases to ensure specification compliance. The process variation problem affects all types of power and timing analysis using the bottom-up characterization approach. With SPICE, the problem is more severe due to the accuracy sought at this level of analysis.

2.2 Discrete Transistor Modeling and Analysis

In SPICE, a transistor is modeled with a set of basic components using mathematical equations. The solution of the node currents and voltages requires complex numerical analysis involving matrix operations. Equation (2.1) gives a simple SPICE transistor model to compute I_{ds} as a function of V_{gs} and V_{ds}. The model is obtained by the first order approximation of the nonlinear equation at the operating point V_{gso} and V_{dso}

$$I_{ds} = f(V_{gs}, V_{ds})$$
$$\approx f(V_{gso}, V_{dso}) + \frac{\partial}{\partial V_{gs}} f(V_{gs0}, V_{dso})\,(V_{gs} - V_{gso})$$
$$+ \frac{\partial}{\partial V_{ds}} f(V_{gso}, V_{dso})\,(V_{ds} - V_{dso}) \tag{2.1}$$

to form a linear approximation equation. In the small signal model, the equation can be simplified as

$$i_{ds} \approx i_0 + g_m v_{gs} + r_{ds} \tag{2.2}$$

which leads to the model shown in Figure 2.2. The linear equation has to be numerically evaluated in SPICE whenever the operating point V_{gso} and V_{dso} changes, resulting in excessive computation requirements.

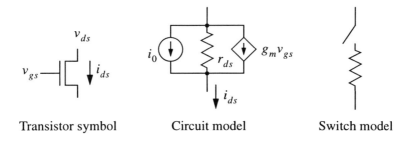

Transistor symbol   Circuit model   Switch model

FIGURE 2.2 Circuit level and switch level representation of a transistor.

2.2.1 Tabular Transistor Model

One way to speed up computation is to express the transistor models in tabular forms stored in a database. Now instead of evaluating equations, a simple table lookup is used. The process involves setting the voltage step increments and running SPICE to compute the current values for the lookup tables. During circuit analysis, a quick table-lookup is used to obtain i_{ds}, given the operating points V_{gso} and V_{dso}.

Circuit level analysis tools based on tabular transistor models have been reported [2.8]. A one-time characterization effort using SPICE simulation of the MOS transistors with various sizes is required to build the lookup tables for the discrete transistor models. The system was mainly designed for timing and power analysis of digital circuits. To speed up computation, it applies the *event-driven* approach, in which an event is registered when a significant change in node voltage occurs. If the event-driven approach fails, it rolls back to the circuit analysis method. A transient analysis method similar to SPICE is used to perform time step simulation. A notorious problem in simulating large circuits with SPICE is the DC convergence problem. Bulk of the simulation time is often attributed to find the initial DC solution of the circuit so that it converges to a legal state. By using the knowledge that the circuit is mainly digital, the logic values from the primary inputs can be propagated to the circuit to help solve the DC convergence problem.

The transistor model quantization process introduces inaccuracies but improves the speed of analysis. The maximum circuit size and analysis speed using the tabular transistor model improves nearly two orders of magnitude compared to SPICE. The accuracy loss is mostly tolerable for digital circuits. Even with the increased capacity, transistor-level tools are often not able to perform full-chip power analysis of very large scale chips. The analysis speed is still several orders of magnitude slower than logic-level simulation. Chips with up to a hundred thousand transistors can be handled by transistor-level tools but many VLSI chips today exceed that scale.

2.2.2 Switch Level Analysis

Most digital circuit analysis is restricted to several basic circuit components such as transistors, capacitors and resistors. Because of the restricted component types, computation speed and memory can be improved by using higher-level abstraction model with little loss in accuracy. One such analysis is called *switch-level* simulation.

The basic idea of the switch-level simulation is to view a transistor as a two-state on-off switch with a resistor, as shown in Figure 2.2. A transistor is turned on when its gate voltage exceeds the threshold voltage. Under this model, timing simulation can be performed using approximated RC calculation that is more efficient than transistor model analysis.

Simulation tools for switch-level timing analysis have been reported [2.9] [2.10]. Recently, power analysis tools based on these simulators have also been developed. The power dissipation is estimated from the switching frequency and capacitance of each node. Short-circuit power can also be accounted by observing the time in which the switches form a power-ground path. The accuracy of switch-level analysis is worse than circuit-level analysis but offers faster speed.

2.3 Gate-level Logic Simulation

Simulation-based gate-level timing analysis has been a very mature technique in today's VLSI design. The component abstraction at this level is logic gates and nets. The circuit consists of components having defined logic behavior at its inputs and outputs, such as NAND gates, latches and flip-flops. Most gate-level analysis can also handle capacitors and some can also handle resistors and restricted models of interconnect wires. Gate-level logic simulation software is one of the earliest CAD tools being developed. Today, many gate-level logic simulators are available, most of which can perform full-chip simulation up to several million gates.

2.3.1 Basics of Gate-level Analysis

The most popular gate-level analysis is based on the so called *event-driven* logic simulation. Events are zero-one logic switching of nets in a circuit at a particular simulation time point. As one switching event occurs at the input of a logic gate, it may trigger other events at the output of the gate after a specified time delay. Computer simulation of such events provides a very accurate pre-fabrication logic analysis and verification of digital chips. Most gate-level simulation also supports other logic states such as, "unknown," "don't care" and "high-impedance," to help the designer to simulate the circuit in a more realistic manner. Some simulators offer an extensive set of logic states for added accuracy in timing analysis. *Verilog* and *VHDL* are two popular languages used to describe gate-level design.

Recently, the *cycle-based* simulators are being introduced into the design community. Such simulators assume that circuits are driven by synchronous master clock signals. Instead of scheduling events at arbitrary time points, certain nets of the circuit are only allowed a handful of events at a given clock cycle. This reduces the number of events to be simulated and results in more efficient analysis.

Many gate-level simulators are so mature that special purpose computer hardware has been used to speed up the simulation algorithms. The idea is similar to the graphic co-processor in a computer system. Instead of using a general purpose CPU to execute the simulation program, special purpose hardware optimized for logic simulation is used. This *hardware acceleration* technology generally results in several factors of speedup compared to using a general purpose computing system.

Another technology that offers several orders of magnitude speedup in gate-level analysis is called *hardware emulation*. Instead of simulating switching events using software programs, the logic network is partitioned into smaller manageable sub-blocks. The Boolean function of each sub-block is extracted and implemented with a hardware table mapping mechanism such as RAM or FPGA. A reconfigurable inter-connection network, carrying the logic signals, binds the sub-blocks together. Circuits up to a million gates can be emulated with this technology but this is also the most expensive type of logic simulator to operate and maintain because of the sophisticated high-speed hardware required. The simulation speed is only one to two orders of magnitude slower than the actual VLSI chips to be fabricated. For example, a 200MHz CPU can be emulated with a 2MHz clock rate, permitting moderate real-time simulation.

2.3.2 Capacitive Power Dissipation

Gate-level power analysis based on logic simulation is one of the earliest power analysis tools developed. The basic principle of such tools is to perform a logic simulation of the gate-level circuit to obtain the switching activity information. The information is then used to derive the power dissipation of the circuit.

A major advantage of gate-level power analysis is that the $P = CV^2f$ equation can be computed precisely and easily. In a non-logic abstraction such as SPICE, the notion of the frequency of a node is not well defined because it has an analog waveform that is potentially non-periodic and non-digital. In logic simulation, the switching activities of each node can be monitored to determine its frequency. The power dissipated due to charging and discharging capacitors can be easily computed. Each net i of a gate-level circuit is associated with a capacitance C_i and a counter variable t_i. As simulation progresses, a logic switching at net i increments the counter t_i. At the end of the simulation, the frequency of net i is given by $f_i = t_i/(2T)$ where T is the simulation time elapsed. The capacitive power dissipation of the circuit is

$$P_{cap} = \sum_{\text{net } i} C_i V^2 f_i \qquad (2.3)$$

The simple gate-level power calculator is very useful in providing a quick estimate of the chip power dissipation.

In the pre-layout phase, the capacitance C_i can be estimated, as will be discussed in Section 2.3.5. After floorplanning, the node capacitance can also be estimated from the partition and placement of the gates. At the post-layout phase, the capacitance of a node can be accurately extracted from the mask geometry. Many commercial CAD tools can perform the layout extraction for power and timing analysis.

2.3.3 Internal Switching Energy

Equation (2.3) only computes the power dissipated due to charging and discharging of node capacitance. If a node appears inside a logic cell, its switching activities are not accounted because the logic-level abstraction does not define internal nodes. Short-circuit power is also not captured by the equation. The dynamic power dissipated inside the logic cell is called *internal power*, which consists of short-circuit power and charging/discharging of internal nodes.

For a simple logic gate, the internal power consumed by the gate can be computed through a *characterization* process similar to that of timing analysis for logic gates [2.3]. The idea is to simulate the "dynamic energy dissipation events" of the gate with SPICE or other lower-level power simulation tools. For example, in a NAND gate with inputs A, B and output Y, the logic event "A = 1, B switches from 0 to 1" causes the output to switch from 1 to 0 and consumes some amount of dynamic energy internally. The energy is caused by short-circuit current or charging/discharging of internal nodes in the gate. The dynamic energy dissipation event can be easily observed during logic simulation.

The computation of dynamic internal power uses the concept of logic events. Each gate has a pre-defined set of logic events in which a quantum of energy is consumed for each event. The energy value for each event can be computed with SPICE circuit simulation. For example, a simple 4-transistor NAND gate has four dynamic energy dissipation events as shown in Figure 2.3(b). The typical energy consumption of each event is also shown in the figure. This energy accounts for the short-circuit current and charging or discharging of internal nodes of the gate. With the energy associated with each event, we only need to know the occurrence frequency of each event from the logic simulation to compute the power dissipation associated with the event. The computation is repeated for all events of all gates in the circuit to obtain the total dynamic internal power dissipation as follows

$$P_{int} = \sum_{\text{gate } g} \sum_{\text{event } e} E(g, e) f(g, e) \qquad (2.4)$$

In the above equation, $E(g, e)$ is the energy of the event e of gate g obtained from logic gate characterization and $f(g, e)$ is the occurrence frequency of the event on the gate observed from logic simulation. The parameter $E(g, e)$ depends on many factors:

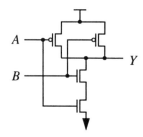

(a) A 4-transistor CMOS NAND gate.

A	B	Y	Dyn energy (pJ)
1	r	f	1.67
1	f	r	1.39
r	1	f	1.94
f	1	r	1.72

A	B	Y	Static power (pW)
0	0	1	5.05
0	1	1	13.1
1	0	1	5.10
1	1	0	28.5

(b) Dynamic energy dissipation events.   (c) Static power dissipation states.

FIGURE 2.3 Dynamic events and static states of a 2-input CMOS NAND gate.

process conditions, operating voltage, temperature, output loading capacitance, input signal slopes, etc.

Note that the dynamic energy dissipation events not only depend on the Boolean function of the gate, but also the implementation of the gate. Figure 2.4 shows two different implementations of a two-input NAND gate. The first implementation has only four energy dissipation events as shown but the second implementation has two additional events due to the switching of its internal nodes. Given a transistor netlist of a logic gate, the task of determining the complete set of dynamic energy dissipation events requires careful consideration to avoid errors in power analysis.

2.3.4 Static State Power

A similar event characterization idea can also be used to compute the static power dissipation of a logic gate. In this case, the power dissipation depends on the *state* of the logic gate. For example, a two-input NAND gate has four distinct states, as shown in

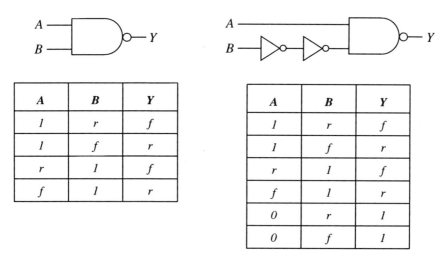

A	B	Y
1	r	f
1	f	r
r	1	f
f	1	r

A	B	Y
1	r	f
1	f	r
r	1	f
f	1	r
0	r	1
0	f	1

FIGURE 2.4 Two different implementations of NAND gate result in different dynamic energy dissipation events.

Figure 2.3(c). Under different states, the transistors operate in different modes and thus the static leakage power of the gate is different. As we have discussed in Section 1.4, the leakage power is primarily determined by the subthreshold and reverse biased leakage of MOS transistors. During logic simulation, we observe the gate for a period T and record the fraction of time $T(g, s)/T$ in which a gate g stays in a particular state s. We perform this observation for all states of the gate to obtain the static leakage of the gate and repeat the computation for all gates to find the total static power P_{stat} as follows

$$P_{stat} = \sum_{\text{gate } g} \sum_{\text{state } s} P(g, s)\frac{T(g, s)}{T} \qquad (2.5)$$

In the above equation, $P(g, s)$ is the static power dissipation of gate g at state s obtained from characterization. The state duration $T(g, s)$ is obtained from logic simulation. It is the total time the gate g stays at state s. The static power $P(g, s)$ depends on process conditions, operating voltage, temperature, etc.

2.3.5 Gate-level Capacitance Estimation

Capacitance is the most important physical attribute that affects the power dissipation of CMOS circuits as evident from Equation (2.3). Capacitance also has a direct impact on delays and signal slopes of logic gates. Changes in gate delays may affect the switching characteristics of the circuit and influence power dissipation. Short-circuit current is affected by the input signal slopes and output capacitance loading (see Section 1.3). Thus, capacitance has a direct and indirect impact on power analysis. The accurate estimation of capacitance is important for power analysis and optimization. Two types of parasitic capacitance exist in CMOS circuits: 1. device parasitic capacitance; 2. wiring capacitance.

The parasitic capacitance of MOS devices can be associated with their terminals. The gate capacitance is heavily dependent on the oxide thickness of the gate that is process dependent. The design dependent factors are the width, length and the shape of the gate. Typically, the shape of a transistor gate is rectangular and the width and length of the gate determine its capacitance. For a gate that "bends," e.g., L-shaped, a correction factor can be used to find its equivalent rectangular width and length. The source and drain capacitance is also estimated from a similar method. The primary capacitance contribution of source and drain terminals is the area and shape of the diffusion regions. In general, a larger transistor has more capacitance in all of its terminals.

In the cell-based design environment, the design and layout of the library cells are made available before the chip design. The capacitance of each pin of a cell is therefore fixed by its circuit and layout. The pin capacitance of a cell can be accurately measured and stored in the cell library. One way to measure the pin capacitance is to use SPICE circuit simulation with the help of the capacitor I-V equation

$$i = C\frac{dv}{dt} \qquad (2.6)$$

We vary the pin voltage ΔV of the cell in time ΔT and observe the current i to obtain the capacitance C. This measurement can be performed during the characterization of the cell.

The second source of parasitic capacitance is wiring capacitance. Wiring capacitance depends on the layer, area and shape of the wire. Typically, the width of routing wires is set to the minimum and the wiring capacitance is estimated from the lengths of the wires. In practice, the process dependent factors of wiring capacitance are expressed by a capacitance-per-unit-length parameter that depends on the thickness of the wire, its distance from the substrate and its width. Once the length of a wire is known, wiring capacitance can be computed.

Since wiring capacitance depends on the placement and routing of the gate-level netlist, accurate estimation cannot be obtained before the physical design phase. However, there is still a need to perform capacitance *estimation* before physical design because the latter are lengthy processes. One way to solve this problem is to predict the wire length of a net from the number of pins incident to the net. This is called the *wire-load* model in ASIC terms. A wire-load model provides a mapping of the net's pin-count to the wiring capacitance without actually knowing the exact length of the net. The mapping table can be constructed from historical data of existing designs. Sometimes, the function also depends on the number of cells of the circuit because the mapping of a ten-thousand-cell module and that of a one-thousand-cell module may be very different. Pre-layout wire-load model coupled with pin capacitance characterization can provide a good capacitance estimate for gate-level power analysis. At the post-layout phase, when the actual lengths of the wires are known, the true wiring capacitance of a net can be used to verify the pre-layout analysis.

2.3.6 Gate-level Power Analysis

The previous sections presented the techniques to obtain the capacitive, internal and static power dissipation of a gate-level circuit using logic simulation. The event-driven gate-level power simulation is summarized as follows:

1. Run logic simulation with a set of input vectors.
2. Monitor the toggle count of each net; obtain capacitive power dissipation P_{cap} with Equation (2.3).
3. Monitor the dynamic energy dissipation events of each gate; obtain internal switching power dissipation P_{int} using Equation (2.4).
4. Monitor the static power dissipation states of each gate; obtain static power dissipation P_{stat} with Equation (2.5).
5. Sum up all power dissipation components.

The total power dissipation of the circuit is the sum of the three power components expressed in Equations (2.3), (2.4) and (2.5)

$$P = P_{cap} + P_{int} + P_{stat} \tag{2.7}$$

The choice of simulation vectors needs to be considered carefully because power dissipation depends on them. The vectors should be chosen such that they reflect the desired operating conditions in which power analysis is sought. Generally, the same vectors used for functional simulation and verification can be used for power analysis. The selection of simulation vectors is application dependent and should be determined from the operating environment of the chip. The vectors for test analysis are obviously not suitable for average power measurement because a chip does not normally operate in the test mode. For example, the simulation vectors of a CPU can be obtained from its instruction trace of standard benchmark software with the proper instruction mix.

The static and internal power dissipation of a gate depends on several factors such as the operating voltage and temperature, output load capacitance, input signal slopes, fabrication process, etc. To capture the power dissipation variation due to such conditions, case analysis can be applied. The gate is simulated with SPICE for all possible conditions that can affect the power. The results are stored in a multi-dimensional table after cells are characterized. During analysis, the actual conditions of the circuit under simulation are specified by the user and the correct internal power or energy values will be used for analysis.

An automated power characterization and analysis tool using this technique has been reported [2.3]. The analysis speed of a gate-level tool is fast enough to allow full-chip simulation. With the static and internal power characterization mentioned above, the accuracy within 10-15% of SPICE simulation is possible. Commercial CAD tools based on this analysis method have been introduced.

Detailed power dissipation of the logic network can be obtained with a simulation technique. For example, we can observe the power dissipation on a per-gate or per-net basis. We can compute the power dissipation of a module to determine the width of the power supply and ground lines to the module. Coupled with layout information, we can plot the temperature profile of the chip to detect regions with potential reliability problems.

A major disadvantage of gate-level analysis is that signal glitches cannot be modeled precisely. Signal glitches are inherently analog phenomena and the simplified zero-one logic model in gate-level analysis fails to capture their effects. The presence of glitches is very sensitive to the signal and gate delays of the circuit. Signal glitches can be a significant source of power dissipation in some VLSI circuits, as such cannot be ignored. However, it is difficult for any analysis model above the logic level to account for the signal glitches precisely. Some attempts to analyze glitches using probabilistic techniques have been reported [2.11]. One technique uses probability to express the presence of glitches but the unpredictable nature of signal glitches prohibits a precise deterministic analysis.

2.4 Architecture-level Analysis

Architecture-level abstraction is the next design abstraction level above logic gates. This is also called block-level or macro-level design. The basic building blocks at this level are registers, adders, multipliers, busses, multiplexors, memories, state machines, etc. Each component performs a dedicated "high-level" function as perceived by the designer.

Today, architecture-level power analysis is becoming more important because more digital circuits are now synthesized from architectural description. As VLSI chips increase in size and complexity, it has become inefficient to design each individual gate. Over the years, the primary design abstraction has moved from the mask, to transistors, to gates and now to the architecture level. The lower level descriptions, such as gates, transistors and masks can now be automatically generated.

The *dynamic event* and *static state* characterization method for logic gates mentioned in Section 2.3.3 and Section 2.3.4 cannot be practically applied to the architectural components because there are too many events and states. Consider a 16-bit adder in which one of the least significant input bits switches from logic 0 to logic 1. Depending on the logic value of the other inputs, the switching may or may not trigger a carry propagation inside the adder. The power dissipation is thus different depending on the logic values of the other inputs. In the worst case, we may need to enumerate $2^{(16+16)}$ (4.29 billion) possible events to fully characterize the 16-bit adder with the gate-level characterization method. The enumeration is finite but certainly not practical to compute.

2.4.1 Power Models Based on Activities

A distinguished feature of most architecture-level components is that they have well-structured regularity. The components are typically constructed by cascading or repeating simpler units built from logic gates. One way to characterize the architectural components is to express the power dissipation as a function of the number of bits of the components and their operating frequencies [2.4]. For example, the power dissipation of an adder can be expressed as

$$P = (nK_1 + K_2)f \qquad (2.8)$$

where n is the number of bits, f is the frequency of the addition operation, K_1 and K_2 are empirical coefficients derived from characterization with a lower-level power analysis such as gate-level simulation. This simple power model depends only on the operating frequency and size of the adder. The model does not take into account the *data dependency* of the power dissipation. For example, if one input of the adder is always zero, we would expect the power dissipation to be less compared with the case when both inputs are changing.

A more accurate model [2.12] that can capture data dependency is to characterize the power dissipation as

$$P = \sum_{\text{input } i} K_i f_i \tag{2.9}$$

for each input signal i of the component. Again we have to perform characterization to derive the coefficients K_i. In some cases, the number of coefficients can be reduced because of the particular characteristics of the component. For example, since addition is commutative, the coefficients for the i-th bit of input A and the i-th bit of input B can be set to the same value with only a slight loss in accuracy. In other components, more K_i's can be set to the same value, such as the coefficients for the data bits of a register file. For larger components with deep logic nesting, e.g., multipliers, the power might also be dependent on the switching activities of the outputs, resulting in a power model as follows

$$P = \sum_{\text{input } i} K_i f_i + \sum_{\text{output } j} K_j f_j \tag{2.10}$$

Equations (2.9) and (2.10) require a factor K_i for each input i. Sometimes this is too tedious to characterize and we can simplify them to

$$P = K_1 \sum_{\text{input } i} f_i + K_2 \sum_{\text{output } j} f_j = K_1 f_{in} + K_2 f_{out} \tag{2.11}$$

The characterization of the coefficients involves generating random signals with the specified frequencies to exercise the architectural components. The power dissipations of the components are observed and the K factors are derived by regression fitting.

2.4.2 Power Model Based on Component Operations

Yet another method to model power dissipation is to express the power in terms of the frequency of some primitive operations of an architecture component. This works well because most architecture-level components only have a few well-defined operations. For example, the power dissipation of a small memory component can be written as

$$P = K_1 f_{read} + K_2 f_{write} \tag{2.12}$$

The parameters f_{read} and f_{write} are the frequencies of READ and WRITE operations, respectively. The parameters can be obtained from simulation and the coefficients K_1 and K_2 are obtained from characterization and properties of the component.

Strictly speaking, the power dissipation of the READ and WRITE operations of a memory component is also dependent on the actual address and data values. However, modeling such fine details of the power dissipation could be prohibitive in terms of computation. The problem is similar to the logic event based modeling in which the enumeration of the logic conditions is too large for our computer to handle. The compromise is to use the average READ and WRITE energy of the operations in Equation (2.12), which introduces some inaccuracies, but improves the computation efficiency and generality of the power model. If the address and data values of the memory operations are fairly random, this solution is often very effective in practice.

In some cases, the memory access pattern is skewed such that most of the READ and WRITE operations occur at a particular location, e.g., address zero. If this is known a priori, the power model can be modified to improve the accuracy. For example, we can characterize the zero address (ZA) and the non-zero address (NZA) operations as

$$P = K_{ZA\text{-}read} f_{ZA\text{-}read} + K_{ZA\text{-}write} f_{ZA\text{-}write}$$
$$+ K_{NZA\text{-}read} f_{NZA\text{-}read} + K_{NZA\text{-}write} f_{NZA\text{-}write} \qquad (2.13)$$

2.4.3 Abstract Statistical Power Models

As we can see from the power models for the various architecture-level components, we have to make certain assumptions on the statistical behaviors of the components in order to model the power dissipation efficiently. The primary reason is that a precise model that characterizes the exact signal events of the architecture-level components is computationally inefficient. The large number of I/O pins of the components prohibits the use of logic events such as the one used for gate-level logic analysis. Power models based on the idea of pin toggling, operations, probabilities and other statistical quantities are used to analyze the power dissipation of the architecture components.

The techniques used for characterizing memory components in Section 2.4.2 can be generalized and formalized to the concept of power models based on abstract statistical parameters of the components. There are several fundamental issues in composing a power model:

1. Which parameters have major influences on the power dissipation? Size, signal frequency, operation frequency, etc.

2. What is the relationship of the power dissipation with the chosen parameters? Equation or lookup table.

The answers to the questions strongly depend on the type and implementation of the components. They also depend on the accuracy level tolerated by the analysis.

Adders, multipliers, memories and registers all have very different operation characteristics and require different parameters for analysis.

In general, the architecture-level power models are not as robust and versatile as the gate-level models because some assumptions have to be made in the modeling and characterization to be computationally efficient. To maintain accuracy, the analysis tools have to observe the operation characteristics of the components to make sure that the assumptions made by the power models are not violated. Nevertheless, the inability to compose accurate power models does not diminish the need for power analysis at the architecture level. A good strategy is to compose power models or analysis methods that emphasize the relative accuracy rather than absolute accuracy. If an architecture design X consumes more power than another design Y, a good power estimation technique should provide such insight; even though the actual power dissipation of designs X and Y is difficult to estimate accurately. As the architecture-level abstraction is the focus of major design decisions, the analysis tools at this level will be used primarily for design trade-off and exploration.

2.5 Data Correlation Analysis in DSP Systems

In a digital signal processing (DSP) system, a phenomenon known as *sample correlation* has been observed. Sample correlation refers to the property that successive data samples are very close in their numerical values and consequently their binary representations have many bits in common. This is not a coincidence but a direct result of sampling a band-limited analog signal with a higher sampling rate relative to the analog signal bandwidth. Some data streams exhibit *negative correlation* (anti-correlation) in which successive samples jump from a large positive value to a large negative value. Negative correlation may occur in some digital signal coding schemes such as delta modulation. Examples of such data streams are depicted in Figure 2.5.

Obviously, positive or negative correlation has a significant effect on the power dissipation of a DSP system because of the switching activities on the system datapath. If we can find the relationship between the data correlation and power dissipation, we can develop a high-level power model without a sample-by-sample analysis of the data stream. Our goal is to estimate power dissipation of an architecture-level component based on the frequency and some correlation measures of the data stream.

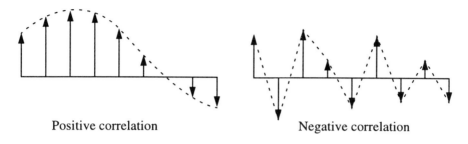

Positive correlation   Negative correlation

FIGURE 2.5 Correlation resulting from sampling analog signals.

2.5.1 Dual Bit Type Signal Model

The effect of data correlation on power dissipation often depends on the numerical representation of the digital system. In digital signal processing, the two's complement representation is the most widely used. Another popular numerical representation method is the signed magnitude. We will assume the use of two's complement representation as the analysis method for signed magnitude representation is similar.

Let us observe the toggle characteristics of the data signals under the influence of data correlation. If the data sample is positively correlated, successive data sample values are very close in their binary representation. This means that the least significant bits (LSB) of the data bus toggle frequently while the most significant bits (MSB) are relatively quiet. If we plot the bit-toggle frequencies of the signals, the characteristics shown in Figure 2.6 will be observed [2.13]. Some of the LSB bits toggle at approximately half the maximum frequency. This is called the *uniform white noise* region because the bits toggle in a random fashion. On the MSB side, the bits have a very low toggle rate and they are called the *sign bit* region. Most of the toggling at this region is the result of a sign change of the data samples, which occurs infrequently. There is also a grey area between the two regions where the toggle frequency changes from white noise to sign bit. In this region, the bit-toggle rate changes from near zero to 0.5, typically in a linear fashion. Note that the switching frequency is normalized with respect to the maximum toggle rate, which is half of the sampling frequency. For a negatively correlated data stream, the converse is observed. The sign bit region has a very high switching frequency and the noise bit region remains at random. If a data stream exhibits no correlation, all bit switching characteristics will all look like uniform white noise.

The above observation allows us to characterize the data stream with only a few parameters:

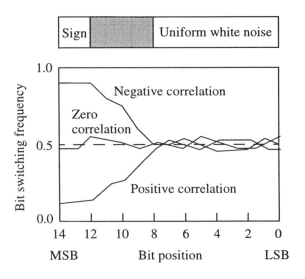

FIGURE 2.6 Effects of data correlation on bit switching frequency.

1. Sample frequency.

2. Data correlation factor from −1.0 to +1.0.

3. The sign bit and uniform white noise regions with two integers.

Such characterization of data signals is called the *dual bit type* model, proposed by Landman and Rabaey [2.13].

2.5.2 Datapath Module Characterization and Power Analysis

The dual bit type signal model provides a very compact representation of the switching characteristics. We like to develop power dissipation models (equations) under such signal excitation so that they can be used for power analysis of architectural components. The power models are sensitive to the signal correlation and the "bit type" of the signals. There are many different methods to characterize the power dissipation and we will describe several methods proposed in [2.13].

For illustration purposes, we first assume that the module to be characterized is a simple single-input single-output block such as a FIFO data queue. We further assume that there is no activity coupling between any two-bit pair so that we can examine a single bit of the component and generalize it to the entire module. We can use a lower-level power analysis tool, such as a gate-level tool, to analyze the power dissipation of a single bit under the uniform white noise signal at a particular frequency f_1 and voltage V_1. Suppose that the power measured by the lower-level tool is P_1. The effective capacitance C_u of the bit is defined as

$$C_u = \frac{P_1}{V_1^2 f_1} \qquad (2.14)$$

The effective capacitance C_u is approximately equal to the capacitance being switched under the white noise signal excitation. Once the effective capacitance is characterized, we can apply it to compute power at a different frequency f_2 and different voltage V_2 under white noise signal excitation

$$P_{noise} = C_u V_2^2 f_2 \qquad (2.15)$$

The concept of effective capacitance can also be used on the module bits under the sign bit signal excitation. However, the effective capacitance in this case is no longer a scalar quantity. Between successive data samples, the sign bit may or may not change sign. Therefore, there are four effective capacitance values: C_{++}, C_{+-}, C_{-+}, C_{--}. The subscript sign pairs represent the changes in sign bit polarity between two consecutive samples. For example, the effective capacitance C_{-+} is the capacitance switched when the first sample is negative and the next sample is positive. In a FIFO data queue, it is most likely that $C_{+-} = C_{-+}$ and $C_{++} = C_{--}$. But in general, we could construct circuits in which all four effective capacitance variables have different values. With the four effective capacitance values characterized by a lower-level power analysis tool, we can construct a power equation similar to Equation (2.15). Let p_{++}, p_{+-}, p_{-+}, p_{--} be the probabilities that sign changes occur in the data stream. The power equation for the bit excitation under the sign bit signal is

$$P_{sign} = (p_{++}C_{++} + p_{+-}C_{+-} + p_{-+}C_{-+} + p_{--}C_{--}) V^2 f \qquad (2.16)$$

where V is the operating voltage and f is the frequency of the data stream. Note that $p_{++} + p_{+-} + p_{-+} + p_{--} = 1$ and for long data sequence $p_{+-} = p_{-+}$ since the number of plus-minus transitions and that of the minus-plus transitions can differ by at most one.

The above discussion has developed the characterization and power evaluation equations for the bit-level power dissipation under white noise and signed signal excitation. For a module that consists of multiple bits, we need to distinguish the white noise bits from the sign bits. This can be done by taking the midpoint of the grey area in Figure 2.6. All bits to the left (right) of the midpoint are considered to have sign bit (white noise) signals. Let N_s (N_u) be the number of bits with sign bit (white noise) signals. The power dissipation P of the module is

$$C_T = N_u C_u + N_s \left(p_{++} C_{++} + p_{+-} C_{+-} + p_{-+} C_{-+} + p_{--} C_{--} \right)$$

$$P = C_T V^2 f \tag{2.17}$$

The effective capacitance C_u, C_{++}, C_{+-}, C_{-+}, C_{--} are obtained during characterization, once per module design. The variables N_s, N_u, p_{++}, p_{+-}, p_{-+}, p_{--}, are obtained from the correlation factor and signal properties of the data stream, typically by analyzing the data stream from the behavior-level simulation. Sometimes the variables can be estimated and supplied by the designer. If the data stream has high positive correlation, we will have $p_{++}, p_{--} \gg p_{+-}, p_{-+}$ and the converse if the data-stream is negatively correlated.

Equation (2.17) assumes that there are no interactions among data bits in the module and allows us to characterization one bit and applies the effective capacitance to the other bits. This is no longer true for modules that have interactions among data bits such as barrel shifters. One way to solve this is to perform the characterization for all possible combinations of N_u and N_s. Since $N_u + N_s$ is equal to the number of bits N of the module, there are only $N + 1$ conditions to be characterized. In this case, Equation (2.17) will be changed to

$$P = C(N_u) V^2 f \tag{2.18}$$

where $C(N_u)$, $0 \le N_u \le N$ is the effective capacitance when there are N_u white noise bits at the module's input.

We now consider modules with two inputs and one output such as adders. The two inputs may have different sign and noise bit separation points. This creates a region at the output of the adder in which the sign and noise bit signals overlap, as shown in Figure 2.7. There are four possible polarity conditions in the sign bit portions of the inputs and output. Therefore, there are $4 \times 4 \times 4 = 64$ possible types of signal transition patterns. A given capacitor in the module may or may not transition under a particular type of transition pattern. This means that there are 64 effective capacitances from $C_{++/++/++}$, $C_{++/++/+-}$ to $C_{--/-/-/--}$. The "u/ss" condition (IN1 has noise bits and IN2 has sign bits) requires another four effective capacitances and so is the "ss/u" condition. The "u/u" input combination only requires one effective capacitance value. In total, there are $64 + 4 + 4 + 1 = 73$ effective capacitance values to be characterized using a lower-level power analysis tool. Note that for an adder, some transition patterns may not occur. For example, $C_{++/++/--}$ is zero because when both inputs have the positive sign, the output cannot have the negative sign. In general, all 73 effective capacitances have to be characterized for an arbitrary two-input one-output module. The power dissipation is given by

$$C_T = N_u C_{u/u} + \sum_{ss\,=\,++} N_{us} C_{u/ss} + \sum_{ss\,=\,++} N_{su} C_{ss/u} + \sum_{ss/ss/ss\,=\,++/++/++} N_s C_{ss/ss/ss}$$

$$P = C_T V^2 f \tag{2.19}$$

The dual bit type characterization method can be generalized to most types of architecture-level components such as multipliers, memories, etc.

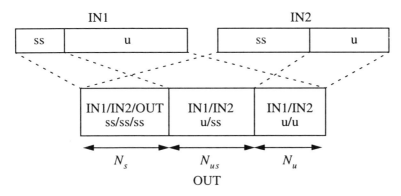

FIGURE 2.7 Signal transition patterns of a two-input, one-output module.

Given the modules characterized by their effective capacitance values, a key step in power analysis is to find the boundary between sign and noise bits. This could be done by simulation or probabilistic analysis. After the "bit type" of each input or output is determined, Equation (2.17) or (2.19) is applied to obtain the power dissipation.

The analysis method is a bottom-up approach meaning that the modules have to be pre-designed and pre-characterized before analysis. This bottom-up approach is acceptable for cell libraries because a cell is small enough to be pre-designed and reused. In general, architecture-level components such as adders, multipliers and register files are seldom pre-designed because the reusability of these components is very limited. Most components have to be designed to fit a specific performance and power constraint. However, once they are characterized, we can analyze the power dissipation of the components under various data correlation and dynamic ranges. Characterization of newly designed components is an inevitable process in any bottom-up approach.

2.6 Monte Carlo Simulation

We have seen various approaches to power analysis based on the dynamic simulation of circuits at different levels of abstraction. All simulation-based power analysis systems require a set of *simulation vectors* at the primary inputs of the circuit to trigger the circuit activities. To obtain the power dissipation of the circuit, the switching activity information is collected and applied to the appropriate power models after simulation. The final power dissipation is subsequently reported from the simulation results. The simulation vectors applied to the circuit have a substantial impact on the final power number reported because power dissipation of a digital circuit heavily depends on its switching activities. Each simulation vector causes some energy to be dissipated and the total power dissipation is derived by summing up the energy of each vector and dividing over the simulation time.

It is crucial that the vectors represent the *typical* conditions at which power estimate is sought. For example, to estimate the typical power dissipation of a CPU chip, the vector set has to be chosen such that it reflects the normal instruction mix of the CPU. We would expect that the CPU consumes more power executing an arithmetic instruction than the "no-operation" instruction. If the simulation vectors do not contain the proper instruction mix, the power analysis result will be skewed. For a CPU, the simulation vectors can be obtained from the instruction trace of executing standard benchmark software.

We will not address the problem of selecting simulation vectors since this depends on the chip and its intended application. Like functional simulation and verification, the chip designer is responsible for providing the proper simulation vectors. Sometimes, randomly generated vectors are sufficient for power analysis but for others a proper vector sequence is required.

Regardless of how the simulation vectors are generated, if we simulate a circuit with only several vectors, we would expect that the power dissipation result obtained is not truthful because the vector length is too short. Most part of the circuit is probably not exercised enough to obtain the actual toggling activities. On the other hand, we can simulate the circuit for millions and millions of vectors to obtain a very accurate measure of the power dissipation. But is it necessary wasting computer resources to simulate that many vectors? How much extra accuracy can we achieve by simulating a million vectors versus only a thousand vectors? How do we know that we have simulated enough vector length? All these lead to an important question of the *stopping criteria* of the simulation: when do we stop simulation so that the result is accurate enough for our purpose?

2.6.1 Statistical Estimation of Mean

The stopping criteria of simulation-based power analysis were first studied by Burch, Najm, Yang and Trick [2.14]. To formulate the problem, we define a basic *sample period* T in which a single power dissipation value is observed. For example, T may be several vectors or several clock cycles. After a particular simulation period T_i, the power dissipation p_i of the circuit during the period T_i is computed. As such, we obtained a series of *power samples* $p_0, p_1, ..., p_N$. The estimated power dissipation P of the circuit under simulation is given by the average value of the samples, i.e.,

$$P = \frac{(p_0 + p_1 + ... + p_N)}{N} \tag{2.20}$$

This is the classical mean estimation problem in statistics in which we draw N samples from a large population and try to determine the mean of the population. With a small value of N, we would expect that the result P is not truthful. If we set N to be too large, unnecessary computation will be performed without gaining meaningful accuracy. So, there is a trade-off between sample size (related to computing efficiency) and accuracy. Our questions of stopping criteria now become the determination of the sample size N.

The power samples p_i are random variables following some unknown probability density function. The distribution of p_i depends on the circuit, simulation vectors and the sample interval. Let μ and σ^2 be the mean and variance of p_i, respectively. The value μ is the true power dissipation of the circuit we are trying to estimate from the N samples of p_i taken. The *sample mean* $P = (p_0 + p_1 + ... + p_N)/N$ is undoubtedly the best estimate of μ we know. The question is: how accurate is P in estimating μ with N samples?

According to the well-known *central limit theorem* in statistics, the sample mean P approaches the normal distribution for large N regardless of the distribution of p_i. For theoretical development, let us assume that the samples p_i have normal distribution. Basic statistical theory states that the average of normally distributed random variables also has normal distribution. The mean of P is exactly μ and its variance is

$$\sigma_P^2 = \frac{\sigma^2}{N} \tag{2.21}$$

As we increase the sample size N, the variance σ_P^2 diminishes so that we obtain more accurate measures of the true mean μ. The normal distribution curve for P is shown in Figure 2.8.

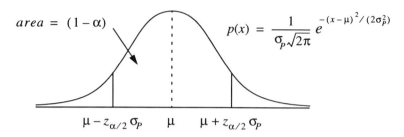

FIGURE 2.8 Normal distribution curve of sample average P.

For the sake of analysis, we assume that μ is positive. To quantify the accuracy of the sample mean P, we prescribe a maximum error tolerance ε, typically with values less than 10%. Given ε, we ask what is the probability that P is within the ε error range of the true mean μ? In other words, what is the probability for the condition $0 \leq |P - \mu|/\mu \leq \varepsilon$? If this probability is high, we will trust the estimate P; otherwise, we must increase the sample size N to gain more confidence. We can obtain this probability by integrating the normal distribution curve $p(x)$ in Figure 2.8. For historical reason, the probability is more conveniently expressed by a confidence variable α. The *confidence level* is defined as $100(1 - \alpha)\%$. A confidence level of 100% ($\alpha = 0$) means that P is absolutely within the error tolerance of ε.[*] Typically, the confidence level is set to more than 90% to be meaningful, i.e., $\alpha \leq 0.1$.

Next, we want to explore the relationships among ε, α and N. We define a variable $z_{\alpha/2}$ such that the area between $\mu - z_{\alpha/2}\sigma_P$ and $\mu + z_{\alpha/2}\sigma_P$ under the normal distribution curve $p(x)$ is $(1 - \alpha)$. From Figure 2.8, this is also the confidence level of the condition $|P - \mu| \leq z_{\alpha/2}\sigma_P$. To ensure the condition $|P - \mu|/\mu \leq \varepsilon$, we require that

$$\frac{|P - \mu|}{\mu} \leq \frac{z_{\alpha/2}\sigma_P}{\mu} \leq \varepsilon \tag{2.22}$$

Substituting Equation (2.21) into the second inequality of (2.22), we have

$$\frac{z_{\alpha/2}\sigma}{\mu\sqrt{N}} \leq \varepsilon \tag{2.23}$$

Since α and ε are fixed constants prescribed by the experimenter, the equation can be rewritten as

---

[*] For any non-zero ε, infinite number of samples are required to achieve 100% confidence.

$$N \geq \left(\frac{z_{\alpha/2}\, \sigma}{\varepsilon\, \mu} \right)^2 \qquad (2.24)$$

Equation (2.24) tells us the minimum number of samples N to be taken in order to be at least $(1 - \alpha)$ confident that the error of P is within the tolerance of ε. In other words, if we make a hundred measurements of P with N samples, only $100\,(1 - \alpha)\%$ of the measurements will have error more than ε.

The value of $z_{\alpha/2}$ is typically obtained from a mathematical table known as the z-distribution function. For most practical purposes, $z_{\alpha/2}$ is a small number. At confidence level of 90% ($\alpha = 0.1$), $z_{0.05}$ is approximately 1.65. At confidence of 99% ($\alpha = 0.01$), $z_{0.005}$ is approximately 2.58. The following table shows some values of $(1 - \alpha)$ and $z_{\alpha/2}$.

TABLE 2.1 **Table of z-distribution.**

$(1 - \alpha)$	$z_{\alpha/2}$
0.9	1.65
0.95	1.96
0.99	2.58
0.998	3.00
1.0	∞

Example:

From experience, the standard deviation of the power samples measured from a circuit has been observed to have ±20% fluctuation from the mean. How many samples are required so that we are 99% confidence that the error of sample mean is within ±5%?

We have $\sigma/\mu = 0.2$, $\varepsilon = 0.05$. Since $(1 - \alpha) = 0.99$, we have $z_{\alpha/2} = 2.58$ from Table 2.1. Therefore

$$N = \left(\frac{z_{\alpha/2}\, \sigma}{\varepsilon\, \mu} \right)^2 = \left(\frac{2.58 \times 0.2}{0.05} \right)^2 = 107 \qquad (2.25)$$

$\square$

Equation (2.24) requires that we know the mean μ and variance σ^2 of the samples p_i. This is actually not very practical because μ and σ^2 are unknown quantities dependent on the circuit, simulation vectors and the sample interval. Since we have taken N samples, $p_0, p_1, ..., p_N$, we can substitute μ with the *sample average* $P = (p_0 + p_1 + ... + p_N)/N$ and σ^2 with the *sample variance* S^2 where

$$S^2 = \frac{1}{N} \sum_{i=1}^{N} (p_i - P)^2 \tag{2.26}$$

The variables P and S^2 are quantities that can be directly computed from the observed N samples of p_i. Since P and S^2 are merely approximations of μ and σ^2, we would expect that more samples need to be taken to achieve the desired confidence level and error tolerance. Indeed this is the case: we only need to change the z-distribution of Equation (2.24) to the t-distribution. Thus to achieve a confidence level of $(1 - \alpha)$ and an error tolerance of ε, the number of samples required is

$$N \geq \left(\frac{t_{[N-1, \alpha/2]} S}{\varepsilon P} \right)^2 \tag{2.27}$$

The quantity $t_{[N-1, \alpha/2]}$ is obtained from the t-distribution function with $N - 1$ degrees of freedom. Again, this quantity is typically taken from a mathematical table. The t-distribution approaches the normal distribution when the degree of freedom increases. For $N > 30$, the t-distribution and the normal distribution are practically indistinguishable. The table of $t_{[N-1, \alpha/2]}$ is shown below:

TABLE 2.2 Table of t-distribution.

N	$(1 - \alpha)$		
	0.9	0.95	0.99
10	1.83	2.26	3.25
20	1.73	2.09	2.86
30	1.70	2.05	2.76
∞	1.65	1.96	2.58

From the above table, we can see that for most practical purposes, $t_{[N-1, \alpha/2]}$ is a small number between 1.5 to 3.5.

2.6.2 Monte Carlo Power Simulation

The last section provides us with the statistical theory to answer the stopping criteria question poses at the end of Section 2.6. As we simulate the circuit, we collect samples p_i at a fixed interval to compute the sample mean $P = (p_0 + p_1 + \ldots + p_N)/N$ and sample variance S^2 using Equation (2.26). Given the confidence level $(1 - \alpha)$ and error tolerance ε, we check if Equation (2.27) is satisfied. If so, we claim that the stopping criteria have been achieved and stop the simulation. This process is called *Monte Carlo power simulation*. The procedure is summarized as follows:

1. Simulate to collect one sample p_i.

2. Evaluate sample mean P and variance S^2 using Equation (2.20) and (2.26).

3. Check if the inequality (2.27) is satisfied; if so stop, else repeat from Step 1.

For computation efficiency, we can replace $t_{[N-1, \alpha/2]}$ with a constant between 1.5 to 3.5, and set a minimum value of N (e.g., 10). We can perform this on any type of simulation as long as we can obtain power samples p_i as we proceed. As more samples are collected, the power dissipation value P becomes more accurate and eventually the stopping criteria are satisfied. Typical plots of the power samples p_i and the sample mean P during simulation are shown in Figure 2.9 and Figure 2.10. From Figure 2.10, we can see that the error of P decreases as more samples are taken over time.

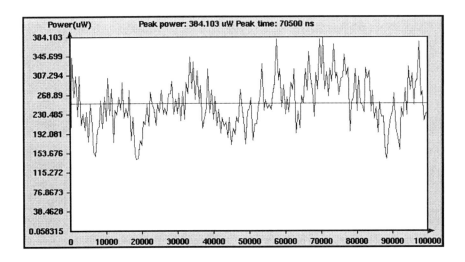

FIGURE 2.9 Graph plot of instantaneous power samples.

Several precautions must be taken in Monte Carlo power simulation. First the samples p_i have to be statistically independent. This is always a fundamental assumption of the statistical analysis. Let us consider a logic-level Monte Carlo simulation. If the

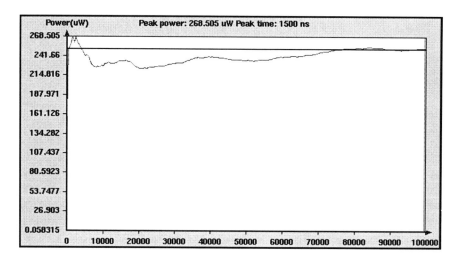

FIGURE 2.10 Graph plot of computed sample mean during simulation.

sample period T is shorter than the typical gate delay, we could violate the independence assumption. This is because when the input of a gate is switching, the output is very likely to switch after the gate delay. Thus, subsequent samples are highly correlated if the sample period is shorter than the gate delay. Of cause, we can always set T to be very large to assure independence but this prolongs the simulation time. As a rule of thumb, we want to set T such that a large number of switching activities occur during the period T to ensure randomness; for examples, several clock cycles. If the signals in the circuit are largely uncorrelated, which is observed to be true for most circuits, the samples p_i should be uncorrelated. We also need to make sure that the simulation vectors do not cause the power samples to become correlated. One way to do this is to pause for a fixed or random amount of time between subsequent samples. As a last resort, if we suspect that the independence assumption is violated, we can always increase the simulation time as a safeguard.

In our theoretical development in the last section, we have assumed that the samples p_i have normal distribution. In general this is difficult to prove for all circuits but there are good reasons to believe that this is true. Note that the power dissipation during a period T is the result of a lot of signal switching in the circuit. If we observe an individual signal of a circuit, its contribution to power dissipation certainly does not follow the normal distribution. However, a power sample p_i is a weighted sum of a large number of signal switches throughout the circuit. We can invoke the central limit theorem to claim that p_i has a normal distribution because it is the sum of a large number of random variables.

Note that the simulation time (related to sample size N) depends on the ratio of the sample variance and sample mean S^2/P^2. One interesting observation is that this ratio typically decreases when the circuit size increases because S^2 is the average of many random variables. When the circuit size increases, the variance of the samples p_i decreases. This reduces N for large circuits due to more uncorrelated nodes. Since it takes more simulation time to compute a power sample p_i, this has an interesting effect that the simulation time is similar for large circuits and small circuits. Intuitively, we can reason that large circuits have more randomness such that subsequent power samples do not vary much. The readers are referred to [2.14] for further theoretical development and experiment results supporting the Monte Carlo simulation approach. A technique for deciding the stopping criteria a priori, i.e., without observing the power samples, has been reported [2.15].

REFERENCES

[2.1]   B. Chowla, H. Gummel and P. Kozah, "MOTIS - an MOS Timing Simulator," *IEEE Transactions on Circuits and Systems*, vol. 22, no. 12, pp. 901-910, Dec. 1975.

[2.2]   W. Eisenmann and H. Graeb, "Fast Transient Power and Noise Estimation for VLSI Circuits," *Proceedings of International Conference of Computer-Aided Design*, pp. 252-257, 1994.

[2.3]   B. George, G. Yeap, M. Wloka, S. Tyler and D. Gossain, "Power Analysis for Semi-Custom Design," *Proceedings of IEEE Custom Integrated Circuits Conference*, pp. 249-252, 1994.

[2.4]   P. Landman and J. Rabaey, "Activity-sensitive Architectural Power Analysis," *IEEE Transactions on Computer-Aided Design of Integrated Circuits and Systems*, vol. 15, no. 6, pp. 571-587, Jun. 1996.

[2.5]   J. Satyanarayana and K. Parhi, "HEAT: Hierarchical Energy Analysis Tool," *Proceedings of Design Automation Conference*, pp. 9-14, 1996.

[2.6]   R. Mehra and J. Rabaey, "Behavioral Level Power Estimation and Exploration," *Proceedings of International Workshop on Low Power Design*, pp. 197-202, 1994.

[2.7]   F. Najm, "A Survey of Power Estimation Techniques in VLSI Circuits," *IEEE Transactions on VLSI Systems*, vol. 2, no. 4, pp. 446-455, Dec. 1994.

[2.8]   C. Huang, B. Zhang, A. Deng and B. Swirski, "The Design and Implementation of PowerMill," *Proceedings of International Symposium on Low Power Design*, pp. 105-109, 1995.

[2.9]   J. Hayes, "An Introduction to Switch-level Modeling," *IEEE Design and Test of Computers*, vol. 4, no. 4. pp. 18-25, Aug. 1987.

[2.10] R. Bryant, "A Switch-level Model and Simulator for MOS Digital Systems." *IEEE Transactions on Computers*, vol. 33, no. 2, pp. 160-177, Feb. 1984.

[2.11] M. Favalli and L. Benini, "Analysis of Glitch Power Dissipation in CMOS ICs," *Proceedings of International Symposium on Low Power Design*, pp. 123-128, 1995.

[2.12] H. Xie, S. Vrudhula, D. Gaitonde, G. Yeap and A. Reyes, "An Efficient Method for Power Characterization of RTL Components Based on a Model of the Power Impact of Input Bits," Technical Report, TR-CLPE-1296-102, University of Arizona.

[2.13] P. Landman and J. Rabaey, "Architectural Power Analysis: The Dual Bit Type Model," *IEEE Transactions on VLSI Systems*, vol. 3, no. 2, pp. 173-187, Jun. 1995.

[2.14] R. Burch, F. Najm, P. Yang and T.N. Trick, "A Monte Carlo Approach for Power Estimation," *IEEE Transactions on VLSI Systems*, vol. 1, no. 11, pp. 63-71, Mar. 1993.

[2.15] A. Hill and S. Kang, "Determining Accuracy Bounds for Simulation-Based Switching Activity Estimation," *IEEE Transactions on Computer-Aided Design of Integrated Circuits and Systems*, vol. 15, no. 6, pp. 611-618, Jun. 1996.

CHAPTER 3

Probabilistic Power Analysis

In Chapter 2, we presented the simulation approach to power analysis. The basic principle of the approach is to mimic the circuit behavior over time, based on a predictable power model. Often the power model is built from a lower-level abstraction process called *characterization*. As we have seen, the simulation approach is very computation intensive. In some cases, we need to solve the voltage or current of the circuit, while in others, we need to simulate the exact time instant on which a certain event occurs in the circuit. In logic simulation, all signal events are recorded precisely by the simulation software. The events are then tabulated and averaged over time to obtain the actual average power dissipation of the circuit.

In this chapter, we introduce a different concept in power analysis, namely *probabilistic analysis*. The basic model of probabilistic power analysis is very different from simulation-based approaches. A logic signal is viewed as a random zero-one process with certain statistical characteristics. We no longer know the exact event time of each logic signal switching. Instead, we only prescribe or derive several numerical statistical characteristics of the signal. The power dissipation of the circuit is then derived from the statistical quantities.

The primary reason for applying probabilistic analysis is computation efficiency. The number of statistical quantities to be analyzed is generally orders of magnitude smaller than the number of events to be processed compared to the simulation approach. Typically, only a few statistical quantities need to be computed at a given node of the circuit as opposed to thousands of events during simulation. The biggest drawback of the probabilistic approach is the loss in accuracy. There are some pathological circuits in which probabilistic analysis results in large error. Nevertheless, probabilistic power estimation technique is a crucial tool in today's digital VLSI design. Techniques that employ combination of simulation and probabilistic analysis have been reported [3.1].

The application of probabilistic power analysis techniques has mainly been developed for gate-level abstraction and above. Since probabilistic techniques are notorious for their preciseness, their applications to transistor or circuit-level networks are limited because of the high accuracy sought at this level. This chapter will focus on probabilistic analysis at the logic level and above.

3.1 Random Logic Signals

The modeling of zero-one logic signals is crucial to the understanding of probabilistic analysis [3.2] [3.3] [3.4]. By capturing only a few essential statistical parameters of a signal, we can construct a very compact description of the signal and analyze its effect on a circuit. Efficient analysis can be performed using this technique. We will discuss some fundamentals of random signal analysis in this section.

3.1.1 Characterization of Logic Signals

A logic signal only consists a waveform with zero-one voltage levels. The most precise way to describe a logic signal is to record all transitions of the signal at the exact times the transitions occur. Consider the logic signal in Figure 3.1. To represent the signal, we write down the initial state of the signal (state 1) and the time value when each transition occurs (5, 15, 20, 35, 45). Such description of the logic waveform allows us to analyze the signals for various purposes. For example, if we wish to compute the frequency of the signal, we count how many times the signal changes state and divide the number by the observation period. This exact characterization of the signal gives us the full details of the signal history, allowing precise reconstruction of the signal. For some purposes, the exact characterization of the signal is too cumbersome, inefficient and results in too much computation resource. For example, if we only wish to know the frequency of the signal, there is no need to know the initial state and the exact switching times; the number of switches should be sufficient.

Consider the two signals shown in Figure 3.2. Although they appear to be different, the number of transitions is identical, which means that the frequencies of both signals are the same. If we wish to apply the $P = CV^2f$ equation, then only the *frequency* of the signal is necessary for the computation. The exact times at which the

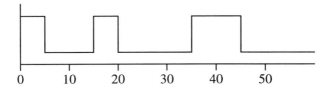

FIGURE 3.1 A logic signal and its characterization.

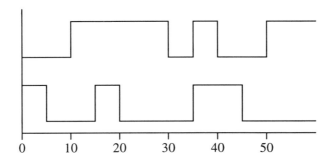

FIGURE 3.2 Two different logic signals with identical frequency.

transitions occur are not important for our computation. All we need to know is the number of transitions per unit time. Thus, for this particular purpose of power dissipation computation, the two signals are indistinguishable. By describing only the frequency of the signal, we can reduce the computation requirements. The disadvantage of this type of signal characterization is that signals are no longer uniquely identified. We only know that a logic signal makes some number of transitions per unit time but the exact instant at which each individual transition occurs is not known.

The above example illustrates an important concept in digital signal analysis. Instead of using the exact historical time domain signal representation, we observe certain *characteristic quantities* (in this case, the frequency) of the signal over a period and use the quantities for power dissipation computation and other analysis. Often, such signal representation allows us to characterize a large number of different signals into a single class. For example, there are infinite number of logic signals with frequency 1MHz. But for the purpose of computing power dissipation using $P = CV^2 f$ equation, all such logic signals result in identical power dissipation.

Many quantities have been proposed and used for this purpose: signal probabilities, correlations, probability density functions, etc. In this chapter, we will discuss the application of such signal representation method in power estimation and analysis of digital VLSI. Sometimes the signal representation may cause accuracy degradation since we have lost the exact details of the signal. The computation efficiency gained from this method can often justify its disadvantages.

3.1.2 Continuous and Discrete Random Signals

The general model for a zero-one logic signal allows a signal transition to occur at any time-point in the real time continuum [3.2]. This is called the *continuous* signal model. The model has the advantage that it can truthfully reflect the precise time of signal transition. However in most digital systems, signal transitions are governed by a system clock signal. All signal transitions occur with respect to the clock signal. Thus, we can restrict the signal to switch only at the multiples of a basic clock period [3.3]. Such a signal model is called the *discrete* model. The discrete model is a special case of the continuous model. It simplifies the representation of a logic signal to a series of zero-one integers and has the advantage of simplicity in signal representation and processing.

3.2 Probability and Frequency

Because of the importance of the $P = CV^2f$ equation in power dissipation analysis, *switching frequency* is a very important characteristic in the analysis of a digital signal. The notion of switching frequency implies an observation period because it is meaningless to talk about the switching frequency of a signal without knowing the observation period. Regardless of the continuous or discrete signal model, the switching frequency f of a digital signal is defined as half the number of transitions per unit time, i.e.,

$$f = \frac{N(T)}{2T} \tag{3.1}$$

where $N(T)$ is the number of logic transitions observed in the period T.

In the continuous random signal model, the observation period is often not specified. Under this situation, it is understood that the observation period is *infinite*. In strict mathematical sense, the period is defined from time zero to T with the value of T approaches infinity

$$f = \lim_{T \to \infty} \frac{N(T)}{2T} \tag{3.2}$$

In the discrete signal model, where a signal is represented by a series of zero-one integers, the basic clock frequency of the signal is always implied and understood. Thus, the actual frequency of the signal in Hz is often normalized with respect to the basic clock rate, yielding a unitless frequency value between zero and one.

3.2.1 Static Probability and Frequency

The *static probability* of a digital signal is the ratio of the time it spends in logic 1 (t_1) to the total observation time $t_0 + t_1$ expressed in a probability value between zero and one, i.e.,

$$p = \frac{t_1}{t_0 + t_1} \tag{3.3}$$

This is a very fundamental characteristic quantity of a logic signal. It states how often the signal stays in logic 1 as opposed to logic 0. Again, when we refer to the static probability of a signal, there is also an implied period of observation. Often the period is infinity if it is not explicitly specified. By definition, the probability that the signal is at logic 0 is $1 - p$.

The static probability and the frequency of a digital signal are related. If the static probability is zero or one, the frequency of the signal has to be zero because if the signal makes a transition, the ratio of logic 1 to logic 0 has to be strictly between zero and one.

Consider a discrete random signal with static probability p. We assume that the signal is *memoryless*, meaning the current state of the logic signal is independent of its history. One way to generate the signal is to toss a biased coin with probability p toward head. If the toss shows head (tail), we set the signal to logic 1 (logic 0). If we perform this experiment many times (e.g., ten thousand tosses), what is the frequency of the discrete random signal?

Consider an experiment of ten thousand tosses. There is a possibility that all ten thousand tosses result in logic 1, which means that the frequency of the signal is zero. There is also the possibility that the tosses yield alternating zeros and ones which results in the maximum frequency of 0.5 (of the basic clock frequency). However, we know that most ten-thousand-toss experiments result in a frequency between the two extremes. What is the value of this frequency? To answer this question, let us review the question of the "frequency" of the random signal. The question is actually not well defined because each of the ten-thousand-toss experiments could give a different frequency number. A more appropriate question to ask is: if we perform infinitely many ten-thousand-toss experiments, what is the *mean* of the frequency value observed? In other words, what is the expected value of the frequency observed in a ten-thousand-toss experiment?

Without involving too much formal mathematics of probability and statistic, we can solve the problem like this: Let's select a state at random. The probability that the state is logic 1 is $p^1 = p$ and the probability that it is logic 0 is $p^0 = (1-p)$. Suppose that the state is logic 1, the *conditional probability* that the next state is also logic 1 is $p^{11} = p$ and the conditional probability that the next state is logic 0 is $p^{10} = (1-p)$. This is a direct consequence of our memoryless assumption. Similarly, we can compute p^{01}, p^{00} and summarize the results as follows:

$$p^0 = p^{00} = p^{10} = (1-p)$$

$$p^1 = p^{01} = p^{11} = p \tag{3.4}$$

The probability T that a transition occurs at a clock boundary is the probability of a zero-to-one transition T^{01} plus the probability of a one-to-zero transition T^{10}

$$T = T^{01} + T^{10} = p^0 p^{01} + p^1 p^{10} = 2p(1-p) \tag{3.5}$$

The expected frequency f is half the transition probability and we have

$$f = p(1-p) \tag{3.6}$$

The above equation relates the static probability p of a memoryless random logic signal to its expected frequency f. Figure 3.3 plots the frequency as a function of the static probability. Note that the frequency is zero at $p = 0$, $p = 1$ and that the maximum expected frequency of 0.25 is achieved when $p = 0.5$. Although statistically the *expected frequency* has the maximum value of 0.25, the *individual* frequency of a particular discrete random signal can still reach the maximum value of 0.5. However, this is highly unlikely as that particular experiment has to yield a long sequence of alternating 0's and 1's.

Because of the memoryless discrete signal model, the static probability is the only independent variable that characterizes the signal. All other properties of the signal, such as the frequency, can be derived from the static probability. For example, to find the average length of a consecutive one's in the zero-one sequence, we note that the probability of observing i consecutive ones followed by a zero is $p^i(1-p)$. Thus, the average length L_1 of consecutive one's is

$$L_1 = \sum_{i=1}^{\infty} i p^i (1-p) = \frac{p}{(1-p)} \tag{3.7}$$

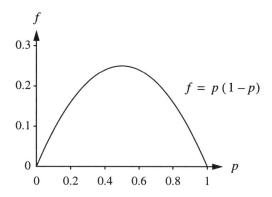

FIGURE 3.3 Expected frequency and static probability of discrete random signals.

3.2.2 Conditional Probability and Frequency

In the memoryless zero-one signal model, the current state of a logic signal is independent of its previous states. We now relax the memoryless condition such that the current state depends on the state immediately preceding it. The generation of such discrete random signals can be done as follows: we prepare two coins, **coin0** and **coin1**. If the previous state is logic 0 (logic 1), we toss **coin0** (**coin1**) to determine the current state. Instead of having one independent variable to characterize the logic signal as in the memoryless case, we now have two independent variables corresponding to the biased probabilities of the two coins. This signal model is called the *lag-one* signal model [3.2] [3.4] [3.5] [3.6].

We define p^{01} (p^{11}) to be the *condition probabilities* that the current state will be logic 1, given that the previous state was logic 0 (logic 1). Notice that p^{01} and p^{11} correspond to the biased probabilities of the two coins described above. Similarly, we define p^{00} (p^{10}) to be the condition probability that the current state is logic 0 given that the previous state was logic 0 (logic 1). The four variables are not independent but related by the following equations

$$p^{01} + p^{00} = 1 \tag{3.8}$$

$$p^{11} + p^{10} = 1 \tag{3.9}$$

Let p^1 be the *static probability* of the random signal as defined in Equation (3.3), i.e., the probability that a logic 1 is observed in any clock cycle. We also define

$$p^0 = 1 - p^1 \tag{3.10}$$

65

as the probability that logic 0 is observed. The static probability $p^1(t)$ of the current state t is dependent on the static probability of the previous state $p^1(t-1)$ by

$$p^1(t) = p^0(t-1)\, p^{01} + p^1(t-1)\, p^{11} \tag{3.11}$$

When the zero-one sequence is time homogeneous (i.e., independent of the choice of time origin), we have $p^1(t) = p^1(t-1)$ and $p^0(t) = p^0(t-1)$. We can drop the time dependency of the static probability and rewrite Equation (3.11) as

$$p^1 = (1-p^1)\, p^{01} + p^1 p^{11} \tag{3.12}$$

which is equivalent to

$$p^1 = \frac{p^{01}}{1-p^{11}+p^{01}} \tag{3.13}$$

Substituting Equation (3.9) into (3.13), we have

$$p^1 = \frac{p^{01}}{p^{10}+p^{01}} \tag{3.14}$$

The equation relates the static probability of the signal to the condition probabilities of transition. When the condition probabilities are given, the static probability of the random signal is determined.

Let us revisit Equations (3.8), (3.9), (3.10) and (3.14) to try to understand their relationships. We have six variables p^{01}, p^{00}, p^{11}, p^{10}, p^1, p^0, each with well-defined physical interpretation. However, the six variables are dependent because they are related by the four equations. There are only two independent variables. When the proper two variables are specified, the other variables can be determined. We can partition the variables into three sets: $\{p^{01}, p^{00}\}$, $\{p^{11}, p^{10}\}$ and $\{p^1, p^0\}$. When any two variables from two *different* sets are specified, the other variables are determined. Thus, only two variables are required to characterize the random signal. Note, however, that certain pairs of values are not possible because they cause the other variables to be out of range, for example, $p^1 = 0.9$ and $p^{10} = 0.5$ forces $p^{01} = 4.5$, which is meaningless.

Since p^{01} and p^{10} completely characterize the signal, the transition probability (and thus frequency) of the signal can be expressed by the two variables. Recall the transition probability of Equation (3.5) and apply Equations (3.10) and (3.14), we have

$$T = p^0 p^{01} + p^1 p^{10} = \frac{2 p^{10} p^{01}}{p^{10} + p^{01}} \tag{3.15}$$

$$f = \frac{T}{2} = \frac{p^{10} p^{01}}{p^{10} + p^{01}} = p^1 p^{10} = p^0 p^{01} \tag{3.16}$$

The maximum frequency $f = 0.5$ is achieved when $p^{10} = p^{01} = 1$, which corresponds to a sequence of alternating 0's and 1's. When $p^{01} = p^{11} = p^1$, we have $p^{10} = p^{00} = p^0 = (1 - p^1)$ and the signal is reduced to the memoryless random signal discussed in the previous section.

The lag-one signal model, with two independent variables, provides one additional degree of freedom over the memoryless signal model that has only one independent variable. Thus, the lag-one model provides a more accurate characterization of random signals. We can generalize the model to "lag-n" with $n + 1$ degree of freedom but the analysis quickly become very complicated.

3.2.3 Word-level and Bit-level Statistics

In a digital signal processing system, the statistical behavior of the data stream is often known at the system design level. The data rate, data stream and sample correlation can be obtained from the system specification of the chip. For example, the voice, image and text processing systems each exhibits certain common characteristics in its input data streams. Also, the observation of the data stream from system-level simulation and analysis can be made to capture the statistical parameters precisely.

The knowledge of the *word-level* data stream statistics [3.7] can become valuable information for architecture-level power estimation. Architecture components such as adders, multipliers and memories operate directly on the word-level data stream. The relationship between the word-level statistics and the *bit-level* statistics is the subject of this section.

The link between the word-level and bit-level statistics is in the data representation. Without loss of generality, we will assume that the word-level data are integers. Let w be an n-bit unsigned integer and let $b_{n-1}, \ldots, b_1, b_0$ be the binary bits of the integer, with b_0 be the least significant bit. The integer value w represented by the binary bits is

$$w = \sum_{i=0}^{n-1} b_i 2^i \tag{3.17}$$

In general, we can write

$$w = \sum_{i=0}^{n-1} b_i k_i \tag{3.18}$$

where $k_0, k_1, \ldots, k_{n-1}$ are some constants derived from the choice of binary representation. For example, in the two's complement representation, we have $k_i = 2^i, 0 \leq i \leq n-2$ and $k_{n-1} = (-2^{n-1})$.

Given Equation (3.18), the static probability p_i (see Section 3.2.1) of bit i is the sum of the probabilities of all words that has $b_i = 1$ in its binary representation, i.e.,

$$p_i = \sum_{\forall \text{ integer } w} P[w|b_i(w) = 1] \tag{3.19}$$

The term $P[w|b_i(w) = 1]$ is the probability of the word w in the data stream where the i-th bit $b_i(w)$ is 1.

The probability $P[w|b_i(w) = 1]$ can assume many different distribution functions and a closed form of the summation in Equation (3.19) may not be possible. Nevertheless, the equation provides the link between the word-level probability and the bit-level probability. When the bit-level probability is known, the switching activities of each bit can be computed using Equation (3.6) or other methods.

Example:

An n-bit unsigned integer representation has $M = 2^n$ words with an *exponential distribution* function where the probability $p(w)$ of word w appearing in the data stream is

$$p(w) = \frac{1}{K} e^{-\lambda w} \tag{3.20}$$

with

$$K = \sum_{w=0}^{M-1} e^{-\lambda w} = \frac{1 - e^{-\lambda M}}{1 - e^{-\lambda}} \tag{3.21}$$

What are the static probability and switching frequency of the least significant bit p_0?

Since the probability distribution function of each data sample is given by Equation (3.20) regardless of its previous values, the bit-level signals are memoryless zero-one random signals. Instead of solving p_0 directly, we will solve $\overline{p_0} = 1 - p_0$, which is the probability of even integer words in unsigned integer representation

$$\overline{p_0} = \sum_{w=0}^{\frac{M}{2}-1} \frac{1}{K} e^{-2\lambda w} = \frac{1 - e^{-\lambda M}}{K(1 - e^{-2\lambda})} = \frac{1 - e^{-\lambda}}{1 - e^{-2\lambda}} \tag{3.22}$$

$$f = p(1-p) = \frac{e^{-\lambda}(1 - e^{-\lambda})^2}{(1 - e^{-2\lambda})^2} \tag{3.23}$$

$\square$

3.3 Probabilistic Power Analysis Techniques

In this section we will present a general power analysis model for logic signals expressed in the random zero-one process. As in the simulation based power analysis, we need a circuit representation. The circuit can be at different abstraction levels such as logic or architecture. Since we no longer have a deterministic signal, an event-driven simulation approach is not possible because we do not know the precise time at which a signal switches. We start with some excitation signals at the primary inputs of the circuit. The random logic signals at the primary inputs are expressed by some statistical quantities. From the primary inputs, we *propagate* the statistical quantities to the internal nodes and outputs of the circuit. The propagation of the statistical quantities is done according to a *probabilistic signal propagation model*. This basic framework for probabilistic power analysis is shown in Figure 3.4. The statistical quantities at the primary inputs $\{p_1, f_1\}$, $\{p_2, f_2\}$, $\{p_3, f_3\}$ are specified. The statistical quantities at the internal nodes and outputs $\{p_i, f_i\}$, $\{p_j, f_j\}$, $\{p_k, f_k\}$ are derived from those of the primary inputs. The power dissipation of the circuit is then computed from the statistical quantities at each node of the circuit, as will be described later.

To reflect the actual circuit behavior, the propagation model should be carefully developed such that if we actually exercise the inputs of the circuit with the specified random signals, the output responses will be as predicted by the propagation model. Since we use a random signal model, there is always an uncertainty factor involved in the analysis. However, a good signal propagation model should predict the response of the circuit with high probability. In some cases, the signal propagation rules can be derived or proven by theoretical analysis. The heart of all probabilistic-based power

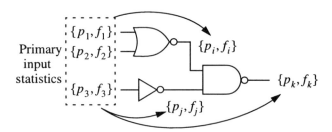

FIGURE 3.4 Propagation of statistical quantities in probabilistic power analysis.

analysis is in the random signal model and its propagation algorithm. We will introduce some basic propagation algorithms later.

An early application of probability propagation analysis to transistor networks for power analysis was reported in [3.8]. The basic idea is to treat each transistor as a switch controlled by its gate signal probability. The signal probability is propagated from the source to the drain of a transistor, modulated by the gate signal probability. In this way, the signal probabilities of all nodes in the circuit are computed and the switching frequencies of the nodes can be derived with Equation (3.6). The power dissipation is subsequently computed with the $P = CV^2f$ equation.

3.3.1 Propagation of Static Probability in Logic Circuits

In Section 3.2.1, we saw the relationship between static probability and frequency under the memoryless random signal model. Since frequency is a crucial parameter in power analysis, the importance of static probability is apparent. If we can find a propagation model for static probability, we can use it to derive the frequency of each node of a circuit, resulting in an efficient power analysis algorithm.

Recall that the static probability of a signal is the probability that the signal is at logic 1 at any given time. The propagation of static probability through a logic gate is quite simple. Consider a two-input AND gate as shown in Figure 3.5. If the static probabilities of the inputs are p_1 and p_2 respectively and the two signals are statistically *uncorrelated*, the output static probability is $p_1 p_2$ because the AND-gate sends out a logic 1 if and only if its inputs are at logic 1. The uncorrelated input signal requirement is crucial for the correctness of this signal propagation model. This independence condition may not be satisfied for all logic gates. However, it is generally assumed that the primary inputs of a circuit have uncorrelated signals. Propagation models that consider signal correlation have been studied [3.4] [3.9] [3.10]. The models have better accuracy but require more computation resources.

FIGURE 3.5 Static probability propagation of a two-input AND gate.

We now derive the general formula for the propagation of static probability through an arbitrary Boolean function. Let $y = f(x_1, ..., x_n)$ be an n-input Boolean function. Applying Shannon's decomposition with respect to x_i, we have

$$y = x_i f_{x_i} + \overline{x_i} f_{\overline{x_i}} \tag{3.24}$$

in which f_{x_i} ($f_{\overline{x_i}}$) is the new Boolean function obtained by setting $x_i = 1$ ($x_i = 0$) in $f(x_1, ..., x_n)$. Let the static probabilities of the input variables be $P(x_1), ..., P(x_n)$. Since the two sum terms in the decomposition cannot be at logic 1 simultaneously, they are mutually exclusive. We can simply add their probabilities

$$P(y) = P(x_i f_{x_i}) + P(\overline{x_i} f_{\overline{x_i}}) = P(x_i)P(f_{x_i}) + P(\overline{x_i})P(f_{\overline{x_i}}) \tag{3.25}$$

The new Boolean functions f_{x_i} and $f_{\overline{x_i}}$ do not contain the variable x_i. The probabilities $P(f_{x_i})$ and $P(f_{\overline{x_i}})$ are computed from the recursive application of Shannon's decomposition to the new Boolean functions. At the end of the recursion, $P(y)$ will be expressed as an arithmetic function of the input probabilities $P(x_i)$. Note that $P(\overline{x_i}) = 1 - P(x_i)$.

Example:

Let $P(a)$, $P(b)$ and $P(c)$ be the input static probabilities. Find the output static probability of $y = ab + c$ by using the Shannon's decomposition method.

$$
\begin{aligned}
P(y) &= P(a)P(b + c) + P(\overline{a})P(c) \\
&= P(a)[P(b) + P(\overline{b})P(c)] + P(\overline{a})P(c) \\
&= P(a)P(b) + P(a)P(c) - P(a)P(b)P(c) + P(c) - P(a)P(c) \\
&= P(a)P(b) + P(c) - P(a)P(b)P(c) \tag{3.26}
\end{aligned}
$$

□

Equation (3.25) gives a method to compute the static probability of an arbitrary Boolean function, given its input static probabilities. This equation serves as a fundamental probability propagation theorem, which is the basis of many other probabilistic power analysis techniques and CAD tools.

The static probability can also be obtained from the truth table of the Boolean function by adding the probabilities of each row where the output is 1. Another method that is generally more computation efficient involves the use of a more elaborate data structure called Boolean Decision Diagram [3.11].

3.3.2 Transition Density Signal Model

Najm [3.12] [3.13] made the first major contribution to the probabilistic analysis of the gate-level circuit. In his formulation, a logic signal is viewed as a zero-one stochastic process characterized by two parameters:

1. Static Probability: the probability that a signal is at logic 1, as defined in Equation (3.3).
2. Transition Density: the number of toggles per unit time.

The random signal model is a generalization of the single number static probability model discussed in Section 3.3.1. Consider two periodic logic signals shown in Figure 3.6. The static probabilities and transition densities of the signals are given in the figure. Note that the top and the middle signals have identical static probabilities but different transition densities. The middle and the bottom signals have identical static probabilities and transition densities and are indistinguishable under this formulation.

The original transition density formulation uses continuous signal model and assumes that the logic signal can make a transition at any point in time. However, the propagation theory can be directly applied to the discrete lag-one signal model discussed in Section 3.2.2. In the lag-one model, when the static probability p and transition density T are specified, the signal is completely characterized and we can use Equation (3.16) to derive p^{10}, p^{01} and other parameters. For example, if given $p^1 = 0.4$ and $T = 0.1$, we can derive $p^{10} = T/(2p^1) = 0.125$ and $p^{01} = T/(2p^0) = 0.083$.

3.3.3 Propagation of Transition Density

Now, given a Boolean function $y = f(x_1, ..., x_n)$, the static probability $P(x_i)$ and transition density $D(x_i)$ of the input variables, we want to find the static probability $P(y)$ and transition density $D(y)$ of the output signal y. The propagation of static probability through an arbitrary Boolean function has already been discussed in Section

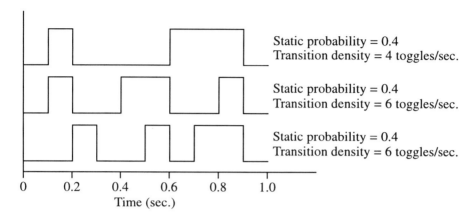

Static probability = 0.4
Transition density = 4 toggles/sec.

Static probability = 0.4
Transition density = 6 toggles/sec.

Static probability = 0.4
Transition density = 6 toggles/sec.

FIGURE 3.6 Static probabilities and transition densities of logic signals.

3.3.1 with the result summarized in Equation (3.25). The same equation is readily applied to the propagation of the static probability under the transition density signal model. So, we will only focus on the propagation of the transition density. Again, we assume that all input signals are uncorrelated.

We will derive the transition density of the output y without going into too much rigorous statistical analysis. We use a *zero gate-delay* model in which the output changes with the input instantaneously. The output y makes a transition because an input x_i switches and the states of the other inputs happen to be such that the transition propagates to the output y. (We assume that there is no simultaneous switching at the input.) If the other inputs are not in the proper states, the transition at x_i will be blocked and does not appear at the output. For example, if one input of an AND-gate is set to zero, the transition at another input will be blocked. Let us try to find the conditions in which an input transition at x_i triggers an output transition.

Recall Shannon's decomposition equation $y = x_i f_{x_i} + \overline{x_i} f_{\overline{x_i}}$ presented in Equation (3.24). As noted before, f_{x_i} ($f_{\overline{x_i}}$) is the new Boolean function obtained by setting $x_i = 1$ ($x_i = 0$) in $f(x_1, ..., x_n)$ and does not contain the variable x_i. Without loss of generality, let's assume that x_i makes a transition from 1 to 0. According to the Shannon's decomposition equation, when $x_i = 1$, the output of y is f_{x_i}. Similarly when $x_i = 0$, the output is $f_{\overline{x_i}}$. If a 1-to-0 or 0-to-1 transition in x_i were to trigger a logic change in y, the Boolean functions f_{x_i} and $f_{\overline{x_i}}$ must have different values. There are only two possible scenarios where this can happen:

1. $f_{x_i} = 1$, $f_{\overline{x_i}} = 0$; or
2. $f_{x_i} = 0$, $f_{\overline{x_i}} = 1$.

73

In other words, the exclusive-OR of the two functions has to be 1, i.e.,

$$f_{x_i} \oplus f_{\bar{x}_i} = 1 \qquad (3.27)$$

The exclusive-OR of the two functions is called the *Boolean difference* of y with respect to x_i, denoted as

$$\frac{dy}{dx_i} = f_{x_i} \oplus f_{\bar{x}_i} \qquad (3.28)$$

The notation dy/dx_i is yet another Boolean function obtained from $f(x_1, \ldots, x_n)$ and does not contain the variable x_i.

From the above discussion, it is clear that an input transition at x_i propagates to the output y if and only if $dy/dx_i = 1$. Let $P(dy/dx_i)$ be the *static probability* that the Boolean function dy/dx_i evaluates to logic 1, and let $D(x_i)$ be the transition density of x_i. Because of the uncorrelated inputs assumption, the output will have transition density of

$$P(\frac{dy}{dx_i}) D(x_i) \qquad (3.29)$$

from the contribution of the signal transition density at the input x_i. We repeat this analysis for all input signals x_i and obtain the total transition density of the output y as

$$D(y) = \sum_{i=1}^{n} P(\frac{dy}{dx_i}) D(x_i) \qquad (3.30)$$

Since dy/dx_i is just another Boolean function, we can compute $P(dy/dx_i)$ given $P(x_i)$'s using Equation (3.25).

Equations (3.25) and (3.30) are the fundamental theorems used to propagate the static probability and transition density through an arbitrary Boolean function $f(x_1, \ldots, x_n)$ under the zero gate-delay model.

Example:

Compute the transition density and static probability of $y = ab + c$ given $P(a) = 0.2$, $P(b) = 0.3$, $P(c) = 0.4$, $D(a) = 1$, $D(b) = 2$, $D(c) = 3$.

From the Example in page 71, the static probability is

$$P(y) = P(a)P(b) + P(c) - P(a)P(b)P(c)$$
$$= 0.2 \times 0.3 + 0.4 - 0.2 \times 0.3 \times 0.4$$
$$= 0.436 \tag{3.31}$$

To compute the transition density, we first find the Boolean difference functions with respect to the inputs

$$\frac{dy}{da} = (b+c) \oplus c = b\,\overline{c}$$

$$\frac{dy}{db} = (a+c) \oplus c = a\,\overline{c}$$

$$\frac{dy}{dc} = 1 \oplus ab = \overline{ab} \tag{3.32}$$

The probabilities of the Boolean difference functions are

$$P(\frac{dy}{da}) = P(b)\,[\,1 - P(c)\,] = 0.18$$

$$P(\frac{dy}{db}) = P(a)\,[\,1 - P(c)\,] = 0.12$$

$$P(\frac{dy}{dc}) = 1 - P(a)P(b) = 0.94 \tag{3.33}$$

and the transition density of the output is

$$D(y) = (0.18 \times 1) + (0.12 \times 2) + (0.94 \times 3) = 3.24 \tag{3.34}$$

$\square$

3.3.4 Gate Level Power Analysis Using Transition Density

Probabilistic analysis provides an alternative to simulation-based approach. The last section presents a general method to propagate transition density through an arbitrary Boolean function. In this section, we discuss the application of the results to probabilistic gate-level power analysis. The basic idea is to propagate the transition densities of the primary inputs to all internal nodes. The transition density values are then used as node frequency and the $P = CV^2f$ equation is applied. The static probabilities and transition densities of the primary inputs are given and assumed to be uncorrelated.

We first present the algorithm for combinational circuits only. The steps are as follows:

1. For each internal node y of the circuit, find the Boolean function of the node with respect to the primary inputs.

2. Find the transition density $D(y)$ of each node $y = f(x_1, \ldots, x_n)$ using Equation (3.30).

3. Compute the total power with the formula $P = \sum\limits_{\text{all nodes } y} 0.5 C_y V^2 D(y)$.

In step 3, $0.5D(y)$ is the frequency and C_y is the capacitance of the node y. The power dissipation is dependent on 1. the input probabilities; 2. transition densities; and 3. the Boolean function of each node with respect to the primary inputs.

The transition density propagation model has been proven by using statistical theory in [3.12]. There is no costly time domain logic simulation and the results have been experimentally observed to be reasonably accurate on combinational circuits. If we generate the random signals according to the probabilities of the inputs and use them to exercise the circuit, the power dissipation observed should be very close to the computed value. The transition density method is more accurate than the pure static probability method introduced in Section 3.3.1 because it uses an additional parameter to characterize the random logic signals. Compared to event-driven logic simulation, it is more computation efficient but produces less accurate power estimate. Commercial CAD software based on this analysis technique has been introduced.

The main disadvantage of the transition density technique is the accuracy. An inherent assumption of the analysis is that the logic gates of the circuits have zero delay. The signal glitches and spurious transitions are not properly modeled. Also, the signal correlations at the primary inputs have been ignored. Extension of the propagation model considering signal correlation issues has been explored [3.4]. However, such extension complicates the propagation algorithm and degrades its computation efficiency.

Perhaps the most serious problem with the transition density analysis is that it only works well on combinational circuits. There is no propagation model of random signals through a sequential cell. Several extensions of the analysis technique to sequential circuits have been reported. The basic principle is to attempt to predict the probabilities and transition densities of the outputs of all sequential elements using some iterative analysis methods. Once the sequential output signal characteristics have been determined, they are propagated to the combinational logic of the circuit. At least two approaches have been proposed to compute the sequential signal probabilities. The analytical approach attempts to approximate the statical quantities of the signals using numerical methods [3.14] [3.15]. Another approach is to perform a high-level simulation of the sequential circuits to determine the sequential probabilities [3.16].

3.4 Signal Entropy

Entropy theory has been successfully used in communication systems to analyze the information contents or capacities of the systems with great success. The application of entropy theory to VLSI power analysis is relatively new. In entropy analysis, the signals in a logic circuit are treated as a collection of random signals. The entropy or randomness of the signals is related to the average switching activities of the circuit. Based on some empirical observations, very high level power models that relate power dissipation to the entropy measures of the systems can be developed.

3.4.1 Basics of Entropy

Entropy is a measure of the randomness carried by a set of discrete events observed over time. Suppose a system has m possible events $\{E_1, E_2, ..., E_m\}$ where each event E_i occurs with probability p_i and $p_1 + p_2 + ... + p_m = 1$. We like to quantify the "information content" carried by an event E_i. If E_i occurs rarely, i.e., $p_i \ll 1$, the information content carried by this particular event should be large because the occurrence of a rare event conveys a lot of messages to us. Conversely if $p_i \approx 1$, it means that the event is almost certain to happen and its information contents should be low. In the studies of information theory, a method to quantify the information content C_i of an event E_i in this manner is to take the logarithm of the event probability

$$C_i = \log_2 \frac{1}{p_i} \tag{3.35}$$

Since $0 \leq p_i \leq 1$, the logarithm term is non-negative and we have $C_i \geq 0$. The quantity C_i is the information content of the event E_i, which is a function of its probability p_i. The unit of this quantity is conveniently defined as *bit*.

The average information content of the *system* is the weighted sum of the information contents of C_i by its occurrence probability p_i. This is also called the *entropy* of the system

$$H = \sum_{i=1}^{m} p_i \log_2 \frac{1}{p_i} \tag{3.36}$$

where H is a positive quantity. The entropy of a system of events can also be interpreted as the *randomness* or the *uncertainty* of the system. The term $p \log_2(1/p)$ is zero at $p = 0$ and $p = 1$. Thus if one of the events of the system $\{E_1, E_2, ..., E_m\}$ has probability one (and other events must have probability zero), the average information content of the system is zero. This matches to our intuition because this is a system with no uncertainty. Equation (3.36) achieves its maximum value of $H = \log_2 m$ if and only if $p_i = 1/m$ for all events.

In logic signal analysis, the events correspond to the word-level values of the signals. In an n-bit logic signal, there are $m = 2^n$ distinct word-level values in the range of $0 \le i \le 2^n - 1$. Let p_i be the probability that the value i occurs. The entropy of the n-bit random logic signal X is

$$H(X) = \sum_{i=0}^{m-1} p_i \log_2 \frac{1}{p_i} \tag{3.37}$$

Sometimes when n is large, computing $H(X)$ is cumbersome because there are too many sum terms. We can assume that each signal bit s_j is independent of the other bits and has static probability of p_j. Each bit is an independent discrete event system with probabilities $\{p_j, (1-p_j)\}$ and we have

$$H(s_j) = p_j \log_2 \frac{1}{p_j} + (1-p_j) \log_2 \frac{1}{(1-p_j)} \tag{3.38}$$

we can approximate $H(X)$ with the sum of the individual bit entropy as follows

$$H(X) = \sum_{j=1}^{n} H(s_j) = \sum_{j=1}^{n} \left[p_j \log_2 \frac{1}{p_j} + (1-p_j) \log_2 \frac{1}{(1-p_j)} \right] \tag{3.39}$$

The maximum entropy of $H(X)$ is $\log_2 m = n$ bits when each bit patterns is equally likely and $p_j = 0.5$ for all j. In general $H(X)$ is low when the probabilities p_j are skewed.

3.4.2 Power Estimation Using Entropy

As discussed in the previous section, the entropy of a set of logic signals is a measure of their "randomness." Intuitively, entropy also correlates to the average switching frequency of the signals. An n-bit signal that rarely toggles suggests that the word-level values are relatively stagnant and many values will probably not appear. This skewed occurrence probability gives a low entropy measure. Conversely if the signal switching is very active, all word-level values are very likely to appear with the same probability and this maximizes the entropy of the signals. Such observation prompts the idea of using signal entropy for power estimation. Below, we present an entropy-based power estimation method proposed by Nemani and Najm [3.17]. Another similar formulation has also been developed [3.18].

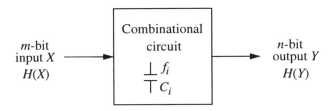

FIGURE 3.7 Power estimation of combinational logic using entropy analysis.

Consider a combinational logic circuit with m-bit input X and n-bit output Y as shown in Figure 3.7. Assuming constant V_{dd}, let f_i be the switching frequency of a node capacitance C_i in the circuit and let N be the total number of nodes. Applying Equation (1.8) in Section 1.2, the power dissipation of the circuit is

$$P = \sum_{i=0}^{N} C_i V_{dd}^2 f_i \tag{3.40}$$

If we assume that $f_i = F$ is constant for all nodes i, we can write

$$P \propto F \sum_{i=0}^{N} C_i = FA \tag{3.41}$$

where A is the sum of node capacitance, proportional to the area or size of the circuit.

Let $H(X)$ and $H(Y)$ be the entropy of the input and output signals respectively as computed by Equation (3.37) or (3.39). From the study of the complexity of Boolean function, a relation has been empirically observed between the average implementation area A and the output entropy $H(Y)$. For small n ($n \leq 10$), the relationship is

$$A \propto 2^n H(Y) \tag{3.42}$$

while for large n,

$$A \propto \frac{2^n}{n} H(Y) \tag{3.43}$$

By making certain assumptions on the structure of the logic circuit, it can also be shown that the average frequency F of the circuit is related to the entropy of the signals

$$F \approx \frac{2}{3(m+n)} [H(X) + H(Y)] \tag{3.44}$$

Substituting Equations (3.43) and (3.44) into Equation (3.41), we obtain

$$P \propto \frac{2^{n+1}}{3n(m+n)} H(Y) [H(X) + H(Y)] \tag{3.45}$$

The proportionality constant depends on the operation voltage and other factors that are largely independent to a particular implementation of the circuit, such as the average node capacitance. Equation (3.45) expresses the power dissipation of a combination circuit with respect to its input and output bit sizes m, n and entropy measures $H(X)$, $H(Y)$ only.

The entropy measures $H(X)$ and $H(Y)$ are typically obtained by monitoring the signals X and Y during a high-level simulation of the circuit. They can also be derived from individual bit probabilities using Equation (3.39). The output entropy $H(Y)$ is dependent on the Boolean functions and the input signals of the circuit. In general, $H(Y) \le H(X)$ because the outputs of a logic gate make fewer transitions than its inputs. It can be theoretically shown that $H(Y) \le H(X)$ for all combinational gates.

The entropy power analysis method conforms to human intuition since circuits with more inputs and outputs or switching activities (expressed by entropy) result in more power dissipation. However, the results should be applied with caution. Equation (3.45) is an *average phenomenon* empirically observed in many digital circuits encountered in VLSI design. Certainly, there are *specific* circuits or classes of circuits that violate the equation; for example, a single CMOS gate. Also, the *absolute accuracy* of the power estimation equation is very poor because there is a lack of implementation details such as voltage, frequency, process technology, devices and cell library. Two identical circuits implemented with different technologies will produce very different absolute power dissipation. When such factors are normalized, the *relative accuracy* of the power estimates is quite suitable for trade-off purposes. For example, if we wish to compare two different state machine implementations M_A and M_B of identical functionality, the power estimation method is applicable. All we need is the number of inputs and outputs of M_A, M_B and their entropy measures. Such comparison is useful to explore early design trade-offs when the implementation details such as the cell library are not available. The lack of absolute accuracy should not discourage its application. Experiment results have shown that the entropy method of power estimation is acceptable for the purpose of high-level power analysis [3.17].

Equation (3.45) is only applicable to large combinational logic circuits with high degree of randomness. The power estimation equation typically fails when the circuits exhibit structural regularity (multipliers, counters and decoders) or highly correlated signals (voice, image). An extension of the analysis to sequential circuits is shown in Figure 3.8. A high-level simulation of the circuit is performed and the input and output entropy measures of the combinational logic are observed. The entropy power estimation method for combinational logic is applied and added to the power dissipation of the sequential elements, which can be easily estimated from the input and output activities of their signals.

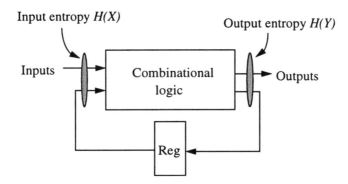

FIGURE 3.8 Entropy analysis of a sequential circuit.

REFERENCES

[3.1]   D. Cheng, K. Cheng, D. Wang and M. Marek-Sadowska, "A New Hybrid Methodology for Power Estimation," *Proceedings of Design Automation Conference*, pp. 439-444, 1996.

[3.2]   S. Vrudhula and H. Xie, "A Fast and Accurate Technique for Estimating Signal Activity in CMOS Logic Circuits," Technical Report, CENG-95-109, University of Arizona, Jan. 1995.

[3.3]   A. Shen, A. Ghosh, S. Devadas and K. Keutzer, "On Average Power Dissipation and Random Pattern Testability of CMOS Combinational Logic Networks," *Proceedings of International Conference of Computer-Aided Design*, pp. 402-407, 1992.

[3.4]   R. Marculescu, D. Marculescu and M. Pedram, "Logic Level Power Estimation Considering Spatiotemporal Correlations," *Proceedings of International Conference of Computer-Aided Design*, pp. 294-299, 1994.

[3.5]   P. Schneider and U. Schlichtmann, "Decomposition of Boolean Functions for Low Power Based on a New Power Estimation Technique," *Proceedings of International Workshop on Low Power Design*, pp. 123-128, 1994.

[3.6]   A. Papoulis, *Probability, Random Variables and Stochastic Processes*, Third Edition, McGraw-Hill, New York, 1991.

[3.7]   S. Ramprasad, N. Shanbhag and I. Hajj, "Estimation of Signal Transition Activity From Word-Level Statistics," Manuscript, 1997.

[3.8]   M. Cirit, "Estimating Dynamic Power Consumption of CMOS Circuits," *Proceedings of International Conference of Computer-Aided Design*, pp. 534-537, 1987.

[3.9]   T. Chou, K. Roy and S. Prasad, "Estimation of Circuit Activity Considering Signal Correlations and Simultaneous Switching," *Proceedings of International Conference of Computer-Aided Design*, pp. 300-303, 1994.

[3.10] P. Schneider and S. Krishnamoorthy, "Effects of Correlations on Accuracy of Power Analysis - An Experimental Study," *Digest of Technical Papers, International Symposium on Low Power Electronics and Design*, pp. 113-116, 1996.

[3.11] R. Bryant, "Graph-based Algorithm for Boolean Function Manipulation," *IEEE Transactions on Computer*, vol. 35, no. 8, pp. 677-691, Aug. 1986.

[3.12] F. Najm, "Transition Density: A New Measure of Activity in Digital Circuits," *IEEE Transactions on Computer-Aided Design of Integrated Circuits and Systems*, vol. 12, no. 2, pp. 310-323, Feb. 1993.

[3.13] F. Najm, "Low-pass Filter for Computing the Transition Density in Digital Circuits," *IEEE Transactions on Computer-Aided Design of Integrated Circuits and Systems*, vol. 13, no. 9, pp. 1123-1131, Sep. 1994.

[3.14] J. Monteiro, S. Devadas and B. Lin, "A Methodology for Efficient Estimation of Switching Activity in Sequential Logic Circuits," *Proceedings of Design Automation Conference*, pp. 12-17, 1994.

[3.15] C. Tsui, M. Pedram and A. Despain, "Exact and Approximate Methods for Calculating Signal and Transition Probabilities in FSMs," *Proceedings of Design Automation Conference*, pp. 18-23, 1994.

[3.16] F. Najm, S. Goel and I. Hajj, "Power Estimation in Sequential Circuits," *Proceedings of Design Automation Conference*, pp. 635-640, 1995.

[3.17] M. Nemani and F. Najm, "Towards a High-Level Power Estimation Capability," *IEEE Transactions on Computer-Aided Design of Integrated Circuits and Systems*, vol. 15, no. 6, pp. 588-598, Jun. 1996.

[3.18] D. Marculescu, R. Marculescu and M. Pedram, "Information Theoretic Measures for Power Analysis," *IEEE Transactions on Computer-Aided Design of Integrated Circuits and Systems*, vol. 15, no. 6, pp. 599-610, Jun. 1996.

| CHAPTER 4 | *Circuit* |

The previous chapters have presented the basic theoretical foundations for the analysis of power dissipation at various levels of design abstraction using different solution methods. From this chapter onwards, we will discuss the optimization and trade-off techniques that involve power dissipation for digital circuits. We begin with the lowest level abstraction and proceed upward. The bottom-up organization of the presentation is intentional because it is important to grasp the problems and solutions at the lower levels before proceeding to the higher levels. Low-level trade-off issues tend to be more rigorous with less uncertainty. The problems can often be formulated precisely and solved as an engineering task. As we move up the abstraction level, the optimization and trade-off problems become less exact and obscured due to more freedom in design configuration and decision. As a result, higher level techniques rely more on human intuition and the art of chip design.

Despite the differences of the optimization and trade-off problems at the various levels of design abstraction, the common themes of the low power techniques are quite similar. It is suggested that the readers revisit Section 1.6 after reading a particular optimization technique to strengthen the fundamentals. This should help the reader in developing the knowledge and skills in low power design.

The power reduction techniques at the circuit-level are quite limited if compared with the other techniques at higher abstraction levels. At the circuit level, percentage power reduction in the teens is considered good. However, circuit techniques can have major impact because some circuits, especially in cell-based design, are repeated thousands of times on a chip. Therefore, circuit techniques with a small percentage improvement should not be overlooked.

Circuits designed manually can often be analyzed in great details. This allows us to optimize the circuit speed, power and area to suit our specification. Many circuits are implemented based on "conventional wisdom" or established practice. Some circuits were designed during the days when power dissipation was not a concern. By redesigning the circuit with the viewpoint of power dissipation, new techniques can often be discovered with good results.

One important circuit technique is the reduction of operating voltage. The general rule is to select the lowest voltage that is acceptable. Often, the implication of operating voltage selection spans beyond the chip design domain to involve fabrication process and system-level considerations. In a complex VLSI design, the choice of operating voltage is seldom made from the viewpoint of a particular abstraction level. We will only discuss voltage selection and trade-off techniques when they appear in narrower context, which makes the problem more tractable.

4.1 Transistor and Gate Sizing

At the circuit-level, transistors are the basic building blocks and a digital circuit can be viewed as a network of transistors with some additional parasitic components such as capacitors and resistors. Having determined the transistor network of a circuit, transistor sizes are the most important factor affecting the quality, i.e., the area, performance and power dissipation, of a circuit. The effect of transistor sizes on a circuit is very complicated because each transistor could potentially introduce one dimension to the problem. In some cases, simple analytical models can be used but there is no analysis model that can be applied to a large class of circuits. Some analyses make strong assumptions on the circuit to facilitate an analytical solution to the sizing problem. One study assumes that the sizing problem is a convex function and Linear Programming can be used to solve the sizing problem optimally [4.1].

Another related problem encountered in cell-based design is gate sizing. Here, the designer has some choices of predefined gates, all with identical logic functionality but different transistor sizes, delays and power dissipation. The goal is to choose a set of gate sizes that best fits the design constraints. We will first present a simple analytic model for the transistor sizing of an inverter chain. Other techniques, practices and CAD tools available for transistor sizing will also be discussed.

4.1.1 Sizing an Inverter Chain

The simplest transistor sizing problem is that of an inverter chain, which has been extensively studied. The general design problem is to drive a large capacitive load without excessive delay, area and power requirements. A large inverter is required to drive the large capacitive load at the final stage. Because the gate capacitance of an inverter is proportional to its size, a medium size inverter is required to drive the large inverter. Therefore, an obvious solution is to use a chain of successively larger inverters so that all inverters in the chain drive appropriate loads and do not cause excessive delay. The circuit structure is shown in Figure 4.1. Intuitively, if the chain is too long, the signal delay will be too large due to the intrinsic delay of each inverter. However, if the chain is too short, the output signal slope will be very weak with long rise and fall times, which again cause long delay. The challenge is to decide the length of the chain, i.e., how many inverters, and the size of each inverter.

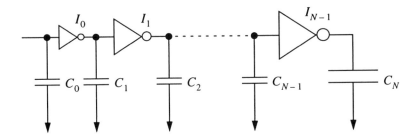

FIGURE 4.1 Sizing an inverter chain.

The problem was first posed in [4.2]. We assume that the P and N transistor size ratio of all inverters are fixed. Typically, they are adjusted such that the rise and fall times are equal. This allows us to use a single number to express the size of an inverter. To simplify the design analysis, we often fix the *stage ratio* of the inverter chain to a constant factor K. This means that the first inverter has a unit size (typically the smallest size), the second has size K, the third has size K^2, etc.

Referring to Figure 4.1, C_i represents the input capacitance of inverter I_i and C_N is the final load capacitance to be driven. The first order analysis of CMOS inverter reveals that the input capacitance of an inverter is proportional to its size. Thus, $C_i / C_{i-1} = K$ and we have $C_N / C_0 = K^N$ or equivalently,

$$N = \frac{\ln(C_N / C_0)}{\ln K} \tag{4.1}$$

Using a very simple RC delay model, the delay of an inverter is directly proportional to the load it drives. In other words, if an inverter drives K other inverters of the same size, the delay through the inverter is Kd where d is the intrinsic delay of the inverter under a single load. Assume that d is identical for all inverters in our inverter chain, the delay D through the chain is

$$D = NKd = \ln\left(\frac{C_N}{C_0}\right)\left(\frac{K}{\ln K}\right)d \tag{4.2}$$

Note that $\ln(C_N/C_0)$ and d are constants and the total chain delay D depends on the stage ratio K. The graph of $K/\ln K$ is shown in Figure 4.2. The delay D is minimum when $K = e = 2.71$, the *natural logarithm base*. Detailed analysis using more elaborate timing model reveals that the optimum stage ratio is between 3 and 5 depending on the fabrication process [4.3].

Let us analyze the power dissipation of the inverter chain. Consider a pair of consecutive inverters in the chain. Since the inverter sizes and their output loading increases K times, the output signal slopes τ of both inverters remain the same. Thus, the duration of short-circuit current for both inverters are identical. This means that the power dissipation of the seceding inverter increases K times compared to that of the preceding inverter. The capacitance charging and discharging power also increases K times because $K = C_i/C_{i-1}$.

Let P_i be the power dissipation of the i-th inverter with input signal slope τ. For $1 \leq i \leq N-1$, we have

$$P_i = KP_{i-1} \tag{4.3}$$

Since C_0 is not driven by the chain, we exclude its switching power. The total power dissipation P of the chain is

$$P = \sum_{i=0}^{N-1} P_i = \sum_{i=0}^{N-1} K^i P_0 = \frac{K^N - 1}{K - 1} P_0 \tag{4.4}$$

The power dissipation of the first inverter P_0 consists of the capacitive power of charging and discharging capacitor C_1 and its short-circuit power when the input slope is τ. Let V and f be the operating voltage and frequency of the inverter chain, respectively. The power P_0 can be expressed as

$$P_0 = C_1 V^2 f + \tau S_0 f = Kf\left(C_0 V^2 + \frac{\tau}{K} S_0\right) \tag{4.5}$$

The term τS_0 is the short-circuit energy of the inverter I_0 per signal transition. The signal slope of the inverter chain τ is directly proportional to the stage ratio K and therefore τ/K is a constant. Since f, C_0, V and S_0 are constants, we have

$$P_0 \propto K \tag{4.6}$$

Combining Equations (4.4), (4.6) and notice that $K^N = C_N/C_0$ is a constant independent of K, we obtain the relationship between the inverter chain power dissipation and the stage ratio

$$P \propto \frac{K}{K-1} \tag{4.7}$$

Equations (4.2) and (4.7) describe the delay and power dissipation of the inverter chain as the stage ratio K varies. There is an optimum delay at $K = 2.71$ but increasing K decreases the power dissipation of the chain. The reason is that increasing K leads to fewer inverters in the chain, less capacitance to be switched and less power dissipation. The graph plot of the inverter chain power versus the stage ratio K is shown in Figure 4.2.

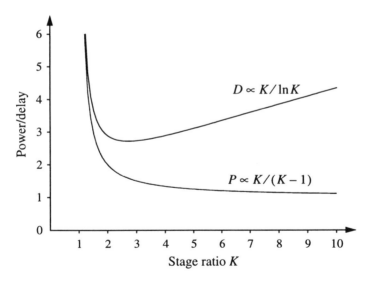

FIGURE 4.2 Graph plot of inverter chain delay $K/\ln K$ and power dissipation $K/(K-1)$.

Having a stage ratio less than the optimum delay point of $K = 2.71$ is not desirable because both delay and power increases sharply. For a stage ratio larger than the optimum, we have the classic power and delay trade-off. At this region, the power decreases but the delay increases. As K increases, the slope of the delay curve becomes steeper but the slope of the power curve becomes flatter. Therefore, setting large values of K far beyond the optimum delay value is also not desirable because the incremental improvement in power diminishes while the delay increases rapidly. For example, from $K = 5$ to $K = 6$, the delay increases 7.2% while the power only improves 4%. In practice, a stage ratio between 3 and 8 is appropriate.

Throughout the above analysis, we have assumed that the stage ratio K is a constant to simplify the analysis. Some studies have shown that using a variable stage ratio of $K(1.2)^i$ can lead to better results [4.4].

4.1.2 Transistor and Gate Sizing for Dynamic Power Reduction

Equations (4.2) and (4.7) illustrate the fundamental trade-off between delay and dynamic power dissipation when driving a large capacitive load. A large gate is required to drive a large load with acceptable delay but requires more power. In many ways, this is similar to the classical delay-area trade-off problem. Therefore if we are not allowed to restructure the transistor network, the sizing for dynamic power reduction generally has the same goal as the area reduction problem. The basic rule is to use the smallest transistors or gates that satisfy the delay constraints [4.5]. To reduce dynamic power, the gates that toggle with higher frequency should be made smaller. Although the basic rule sounds simple, the actual sizing problem is very complicated. Consider part of a circuit as shown in Figure 4.3. Suppose that the gates are not on the critical delay path and should be sized down. We can size down the first gate, the second gate, or both, subject to the available sizes in the cell library as long as the path delay is not violated. Each choice presents a different area-power cost and it is not immediately obvious that which choice leads to the best result. If the path contains many gates, the optimization problem quickly becomes very complicated.

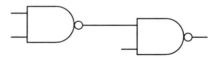

FIGURE 4.3 Gate sizing problem.

The concept of *slack time* is often used to express the timing constraints of the circuit. The slack time of a gate is the difference between the signal required time and the signal arrival time at the output of a gate. A positive slack time means that the signal arrived earlier than its required time and the gate can be sized down. The goal of gate sizing is to adjust the gate sizes such that the slack time of each gate is as low as possible without any gate having a negative slack, i.e., timing violation. In [4.6], a greedy algorithm was proposed. It attempts to downsize a set of gates that give the most reduction in power dissipation without affecting the timing slack of all other gates. The new slack times of the downsized gates are then computed and a new set of gates is selected.

The area minimum sizing problem has been a subject of research in logic synthesis for dozens of years. Today in a top-down cell-based design environment, gate sizing has been very much automated by the logic synthesis system. In the past, only the delay and area trade-off is considered by the logic synthesis system. Recently, power dissipation has been added into the optimization criteria. The added dimension of power dissipation in the optimization increases the complexity of the problem because the area, delay and power cost functions interact in a very complicated manner. Nevertheless, commercial logic synthesis tools that can perform trade-offs between power, delay and area have been introduced. Therefore, the designer is seldom asked to perform transistor or gate sizing manually. The in-depth treatment of automated gate sizing for power dissipation is beyond the scope of this book.

4.1.3 Transistor Sizing for Leakage Power Reduction

An interesting problem occurs when the sizing goal is to reduce the leakage power of a circuit. In Section 1.4, we have noticed that the leakage current of a transistor increases with decreasing threshold voltage and channel length. In general, a lower threshold or shorter channel transistor can provide more saturation current and thus offers a faster transistor. This presents a trade-off between leakage power and delay.

The leakage power of a digital circuit depends on the logic state of the circuit. Consider a simple two-transistor inverter. If the output of the inverter is at logic high, the leakage current of the inverter is determined by the N-transistor that is turned off. Conversely if the output is low, the leakage current depends on the P-transistor. Suppose that the inverter output is mostly at logic high, for example the circuit has a dormant mode in which the inverter output is high for a long time. In order to suppress the leakage current, we can increase the threshold voltage or the channel length of the N-transistor. However by doing so, we also increase the delay of the inverter because the N-transistor now offers less saturation current when it is turned on. A closer look reveals that only the *falling* transition delay of the inverter is affect. The *rise* time of the inverter remains the same because the P-transistor size is unchanged. Thus, if we are fortunate that the falling transition of the inverter is not the critical delay (but the rising is), we can use this skewed transistor sizing method to reduce the leakage power without incurring any delay penalty.

This transistor sizing concept is illustrated in Figure 4.4. The inverter at the top is the reference design, with its transistor sizes, rise/fall delay and leakage power noted in the figure. The output of the inverter has a static probability of 0.99, which means that the leakage current is mainly determined by the N-transistor. Suppose that the output rising delay T_{rise} is critical. If we increase the channel length of the N-transistor, as shown by the bottom-left inverter, the leakage current improves without affecting the critical delay T_{rise} but the non-critical delay T_{fall} worsens. On the other hand, we can decrease the channel length of the P transistor to improve the critical delay without affecting the leakage.

Instead of changing the transistor channel length, we can also change the transistor threshold voltage to achieve a similar effect. We have the choice of fast transistors with low threshold voltage and high leakage current (when turned off) or slow transistors with high threshold voltage and low leakage current. By analyzing the rising and falling delay requirements of a gate, we can use asymmetric P and N-transistor sizes either to reduce leakage or to increase speed without penalties.

An important note is that the logic state of the gate should be *skewed* such that the leakage power is dominated by one type of transistor. Otherwise, if the static probability of the output is 0.5, the leakage power is the average of the P and N-transistor leakage current and we have to trade speed for leakage power and vice versa. This technique has been used in aggressive low voltage and low power technology which offers transistors with more than one threshold voltage. The sizing technique is very important for performance critical circuits with prolonged hibernation mode in which leakage power dissipation is a concern. For high performance circuits with dynamic logic, this technique has also been used to reduce the charge leakage caused by sub-threshold conduction, as reported in [4.7].

4.2 Equivalent Pin Ordering

Most combinational digital logic gates found in a cell library have input pins that are logically equivalent. Examples of such gates are ANDs, ORs, XORs and AOIs. Such gates are used frequently because they are natural to the human thinking process. As for circuit implementation, the gates are robust and easy to design. Logically equivalent pins may not have identical circuit characteristics, which means that the pins

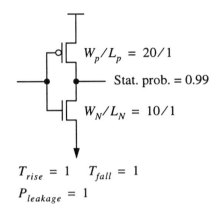

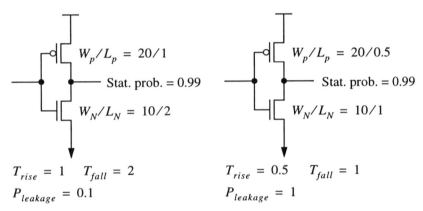

FIGURE 4.4 Transistor sizing for leakage power reduction or speed increase.

have different delay or power consumption. Such property can be exploited for low power design.

Consider a simple two-input CMOS NAND gate shown in Figure 4.5. We examine the condition when the input A is at logic high and the input B switches from logic low to high. Initially, capacitance C_{out} and C_i are charged to V_{dd} because the N-transistor of input A is turned on and the output is at high. After the input B switches and all signals have stabilized, C_{out} and C_i are fully discharged to the ground. The charge stored in C_{out} and C_i has to pass through the N-transistor of input B to discharge. Consider another symmetrical situation in which the input B is at logic high and the input A switches from low to high. Before A switches, C_{out} is charged to V_{dd}

but C_i is at the ground level because input B is high. Thus, less charge transfer is required to complete the logic switching and the delay and power dissipation are less. This simple qualitative analysis illustrates the differences in circuit characteristics of logically equivalent pins. The difference in power dissipation varies depending on various factors such as capacitances and transistor sizes. To conserve power, the inputs should be connected such that transitions from input A to OUT occur more frequently than transitions from input B to OUT. This low power technique is known as *pin ordering.*

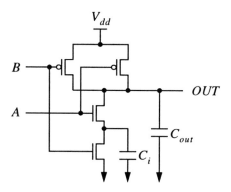

FIGURE 4.5 A two-input CMOS NAND gate.

To apply pin reordering, the designer needs to know the signal statistics at the gate inputs such as the switching frequencies and static probabilities. This can be observed from the logic simulation of a gate-level circuit. Without the probability information, the pin ordering technique cannot be applied. When performing pin reordering, the timing constraints of the design should not be violated. In the NAND gate example above, the pin with less propagation delay (input A) happens to consume less dynamic power than the other input. For other gates, a low power pin may not coincide with low delay and we have to consider the trade-off. In CMOS combinational gates, an input transition involving transistors closer to the output results in better delay and power dissipation. However, in an actual circuit, other factors such as parasitic capacitance, resistance, transistor sizes and input signals may contribute to the differences in pin characteristics. Observing the differences qualitatively may not be easy, but we can always resort to circuit simulation to measure the differences in timing and power.

4.3 Network Restructuring and Reorganization

The pin reordering technique discussed in the previous section is a special of case of a more general method called *transistor restructuring*. In this method, we restructure the transistors of a combinational cell, based on signal probabilities, to achieve better power efficiency within the allowable timing constraints [4.8]. Various versions of transistor reordering techniques will be discussed in this section.

4.3.1 Transistor Network Restructuring

In CMOS logic design, there is a well-known technique in which a Boolean function composed of AND and OR operators is directly mapped to a complex transistor network that implements the function. The mapping steps are as follows:

1. Each variable in the Boolean function corresponds to a pair of P and N-transistors.

2. For the N-transistor network, an AND operator corresponds to a serial connection and an OR operator corresponds to a parallel connection.

3. For the P transistor network, the composition rule is inverted.

4. An inverter is optionally added to the output of the complex gate to maintain the proper polarity or to ensure signal strength.

For example to compute the Boolean function $Y = A(B + C)$, we serialize the connection between the N-transistor of variable A and the network of $(B + C)$, which is a parallel connection of two N-transistors. The P network is similarly constructed with the serial and parallel rule interchanged. The final circuit is shown in Figure 4.6.

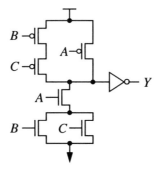

FIGURE 4.6 Network composition of CMOS complex logic gate $Y = A(B + C)$.

The above network construction algorithm does not produce a unique design. Consider an N-transistor network. When an AND operator calls for a serial connection, the ordering of the two sub-networks is irrelevant to the logical behavior of the gate. However, the circuit characteristics depend on the ordering of the serial connection. For example, another implementation of the function $Y = A(B + C)$ is shown in Figure 4.7. For this particular Boolean function, there are two different implementations for each of the P and N networks. Therefore there are four different circuit implementations that are logically equivalent, as shown in Figure 4.8, but different in circuit characteristics such as propagation delay and power dissipation.

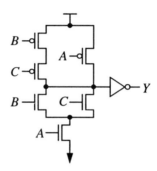

FIGURE 4.7 Alternate circuit implementation of $Y = A(B + C)$.

Based on the above method of reordering a serial network, we can enumerate all equivalent logic networks of a given Boolean function. Sometimes, the Boolean function itself has several equivalent forms and presents more implementation choices. For example, the functions $AB + \overline{A}\,\overline{B} + AC$ and $AB + \overline{A}\,\overline{B} + \overline{B}C$ are equivalent and this can compound to the enumeration of equivalent logic networks.

The analysis of the timing and power efficiency of the different circuit implementations is not a straightforward task. The timing and power characteristics depend on the layout of the cell. As a general rule, transitions involving transistors closer to the output node have less delay and consume less energy. Therefore, the transition probabilities of each input need to be known to evaluate the circuits. In [4.9], a switch level simulation is used to choose the best circuit implementation. Power reduction up to 20% was reported using the transistor network restructuring technique.

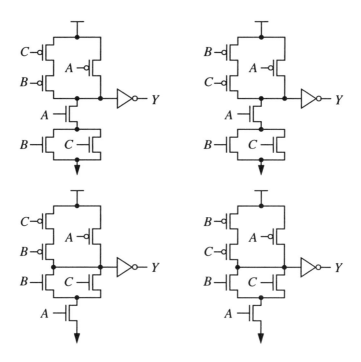

FIGURE 4.8 Four different circuit implementations of $Y = A(B + C)$.

4.3.2 Transistor Network Partitioning and Reorganization

In the last section, we discussed the transistor restructuring technique of a CMOS complex gate. The idea is based on enumerating different circuit implementation of the transistor networks for a Boolean function with AND and OR operators. The restructuring operation is applied to a single CMOS complex gate. In this section, we look beyond a CMOS gate and consider a transistor network. We study the problem of partitioning and reorganizing the network to explore the different power-area-delay trade-off. Network *reorganization* by definition is the task of composing different transistor networks that can implement the same functionality. The problem is so general that it is more of a design art rather than an exact science. The complexity of the problem is equivalent to that of designing a circuit from the specification.

When a Boolean function is large, it cannot be implemented in a single complex gate. One reason is that the number of serial transistor connections is limited to ensure proper operation of the circuit. When transistors are connected in series, the effective resistance of the serial transistor chain increases. To compensate for the increased resistance, the sizes of the transistors have to be increased to maintain an acceptable delay. Also, there is a second order phenomenon known as the *body effect* which increases the effective threshold voltage of the serial transistors, further limiting their conductance. The same problem occurs with too many parallel transistors because each transistor contributes some drain diffusion capacitance to the output node, causing slow transition time. Therefore, a single CMOS gate cannot have too many serial or parallel transistors due to the efficiency concerns. The exact limit of the serial or parallel transistor length is dependent on the process technology, operating voltage, system speed and other factors. As a rule of thumb, the serial chain seldom exceeds 4 to 6 and the parallel transistors are restricted from 8 to 10 in today's CMOS digital circuits.

Given an arbitrary Boolean function described in AND/OR operators, there is an interesting problem of partitioning and reorganizing the circuit into CMOS gates under the serial chain constraint. For example, Figure 4.9(a) and (b) show two ways to implement a 4-input AND operation with a serial chain limit of three. The partition in (a) requires 12 transistors while the partition in (b) requires 14 transistors. Although both circuit organizations implement an identical Boolean function, their circuit characteristics are very different. The circuit in Figure 4.9(a) probably has better area and longest pin-to-pin delay but the other circuit may have better shortest-delay and power dissipation in some situations.

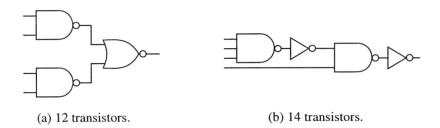

(a) 12 transistors.   (b) 14 transistors.

FIGURE 4.9 Partitioning of a 4-input AND gate with a serial chain limit of three.

Transistor reorganization is a very involved technique because there are many different network structures that can implement the same circuit functionality. The choice of network structures increases exponentially with the circuit size. A human designer, guided by intuition, can only handle small circuits with a limited set of constraints. For large networks, an automated CAD tool has to be used.

A CMOS complex gate reorganization system has been reported [4.10] to improve the power efficiency of combinational digital logic beyond the semicustom standard cell design approach. The design system applied an enumeration process to generate optimized organizations of CMOS complex gates given the Boolean function of the circuit. The circuit partitioning and reorganization procedure is followed by transistor sizing to further improve the circuit performance.

The transistor reorganization technique requires sophisticated physical design technology that is not widely available in the industry. The CMOS complex gates generated by reorganization cannot be pre-designed because there are too many different gate topologies. Furthermore, sizing each individual transistor of a gate requires a tool that can synthesize the transistor layout automatically. The analysis and verification of timing and power have to be performed at the transistor level that is less efficient than at the gate level. The performance and power efficiency is obtained at the cost of complex design and verification processes.

4.4 Special Latches and Flip-flops

Flip-flops and latches are some of the most frequently used elements in digital VLSI. In synchronous systems, they are the starting and ending points of signal delay paths, which decide the maximum speed of the systems. Typically, they consume a large amount of power because they are clocked at the system operating frequency. Careful design of the flip-flop and latch circuits is important to a low power VLSI system. The clock power of a high-performance processor dominates the total power dissipation. In a commercial RISC CPU running at 200MHz, the clock power represents about 40% of the total power consumption [4.11]. A large portion of the clock power is used to drive sequential elements such as flip-flops and latches. Reducing the clock power dissipation of flip-flops and latches is thus an important technique for the total chip power conservation.

The energy dissipation of a flip-flop can be divided into two components: 1. clock energy 2. data energy. The first component is the energy dissipated when the flip-flop is clocked while the data of the flip-flop is unchanged. The second component is the additional energy required to write a different data value into the flip-flop. In a typical flip-flop, the two energy components are comparable. However, in most systems, the data rate of a flip-flop is typically much lower than its clock rate. This means that identical data value is being loaded with very high probability. Thus, the power saving techniques for flip-flops mostly concentrate on the clock energy reduction.

Many different circuit techniques for latches and flip-flops have been explored for various needs. The low power requirement merely adds another factor to the trade-off during the design process. Referring to the $P = CV^2f$ power equation, the techniques for low power flip-flops are obvious. We will not discuss voltage reduction techniques because they are fairly straightforward when applied to flip-flop circuits. Some techniques attempt to reduce the gate capacitance connected to the clock signal. Others attempt to reduce or increase the transistor count to reduce power by minimizing unnecessary internal node switching.

4.4.1 Flip-Flop and Latch Circuits

Figure 4.10 shows various implementations of CMOS latches. Each circuit provides a different trade-off among setup time, hold time, data-to-output and clock-to-output delay. The use NMOS pass transistors instead of transmission gates reduces the loading capacitance of the clock pin at the cost of reduced speed. This eliminates the need for a two-phase non-overlapping clock on the system or a phase splitter inverter that consumes power inside the cell. The circuit suffers from threshold voltage loss when logic 1 is propagated through the NMOS pass transistor. The *single phase latch* circuit shown on the right of Figure 4.10(a) was introduced in [4.12]. It avoids the threshold voltage problem but relies on charge storage effect to retain its data value. This causes some loss in the noise margin but the circuit has been successfully used in high-performance processor design [4.11].

Combined with transistor sizing, circuit-level simulation (e.g., SPICE) can be performed to determine the best circuit implementation for the flip-flops and latches. Process conditions may change the power dissipation characteristics of the devices. Thus, the circuit configuration is not the only factor that affects the switching energy and no general conclusion can be inferred from the structure of the circuit.

Figure 4.11 shows some circuit configurations of flip-flops. A flip-flop is typically constructed from two latches connected together. Note that the single phase dynamic flip-flop at the top-left is a cascaded version of two single phase latches. It is suitable for some low power applications because it does not require internal phase splitting inverter for the clock pin. The circuit at the bottom was reported to achieve lower power at the same speed, with more transistors, compared to a standard flip-flop design on the top-right [4.13].

One circuit uses an exotic differential signaling method to increase speed, at the expense of area and power, as shown in Figure 4.12. This circuit was reportedly used in a RISC microprocessor [4.14]. The flip-flop contains a simple differential feedback circuit that drives an RS latch. The transistors M_1, M_2, M_3, M_4 form a pair of cross coupled feedback inverters. The inputs of the RS latch are precharged and selectively

100

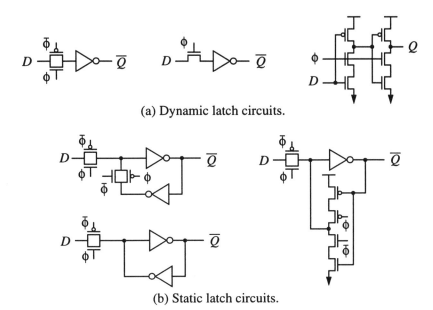

(a) Dynamic latch circuits.

(b) Static latch circuits.

FIGURE 4.10 Various circuit implementations of latches.

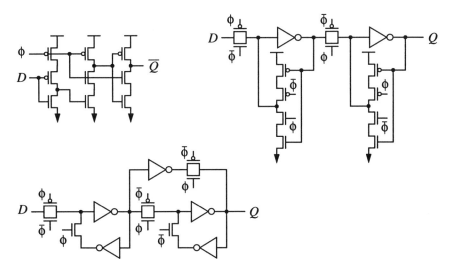

FIGURE 4.11 Flip-flop circuits.

discharged at the rising edge of the clock signal. The RS latch retains the data output during the precharge period when *CLK* is low. The transistor M_5 provides a current path to ground if the input switches after the rising edge of the clock. This prevents the sources of M_2 and M_4 to become floating, which can cause the inputs of the RS latch to have an intermediate voltage level. Although only three transistors are connected to the single phase clock signal, one of the nodes at the inputs of the RS latch is always charged and discharged during evaluation. Note that the full CMOS implementation does not consume static current when the clock is stopped.

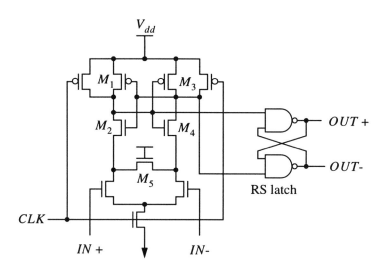

FIGURE 4.12 Differential input latch for high-speed low voltage application.

4.4.2 Self-gating Flip-flop

Part of the clock energy of a flip-flop is consumed by the internal clock buffer to control the transmission gates. If the input of the flip-flop is identical to its output, the switching of its clock signal can be suppressed to conserve power. This idea is similar to the clock gating technique in Section 5.2 and Section 6.1.1. The difference is that the gating function is derived within the flip-flop without any external control signal. Power is saved by holding the internal clock signals of the flip-flop when allowed. The external clock signal of the flip-flop still switches.

An example of the self-gating flip-flop circuit [4.15] is shown in Figure 4.13. The functional behavior of the flip-flop is identical to that of a normal positive edge triggered flip-flop. The transmission gate at *CLK* is used to gate the external clock signal so that the internal clock signals ϕ and $\bar{\phi}$ do not switch if not required. When the D and Q values of the flip-flop are different, the exclusive OR gate output is 1 to ensure that the CLK is passed through. After the signal ϕ turns high, the transmission gate is turned off to avoid further switching unless the D and Q values differ.

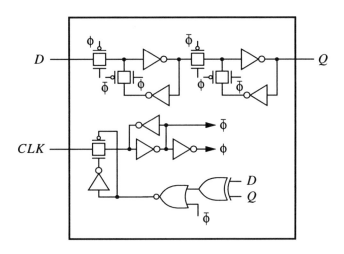

FIGURE 4.13 Flip-flop with self clock-gating.

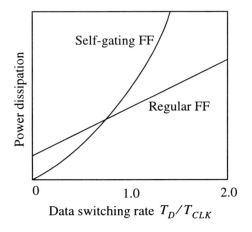

FIGURE 4.14 Power dissipation of self-gating flip-flop and regular flip-flop.

The circuit incurs significant area overhead in the number of transistors. The delay of the flip-flop also degrades slightly because of a longer path from *CLK* to *Q*. If the input switching rate of the flip-flop is sufficiently low compared to the clock rate, the probability of disabling the clock will be high enough to justify the additional power consumed by the comparator and clock gating circuit. The energy efficiency of the self-gating flip-flop compared to the regular flip-flop is shown in Figure 4.14. The power dissipation depends on the relative transition frequencies of the data input T_D and clock T_{CLK}. When $T_D/T_{CLK} \ll 1$, the self-gating flip-flop consumes less power than the regular flip-flop because the internal clock signals can be disabled frequently. When the ratio exceeds a certain value less than 1.0, the self-gating flip-flop will consume more power. The threshold point where the two curves in Figure 4.14 crossed depends on the implementation of the gating circuit. At $T_D/T_{CLK} = 0$, the self-gating flip-flop consumes zero dynamic power because none of its internal node switches but the regular flip-flop consumes power due to its internal clock buffers. Note that if the data input signal has many glitches, the ratio T_D/T_{CLK} becomes very high and the self-gating flip-flop is not suitable.

4.4.3 Combinational Flip-flop

One way to reduce circuit size is to associate logic gates with a flip-flop to produce a combinational flip-flop [4.11]. Two circuits using this technique are shown in Figure 4.15. The multiplexor flip-flop shown at the bottom is often used in systems that employ scan test methodology. Combinational flip-flops are efficient because they are able to eliminate or share redundant gates. The layout of a combinational flip-flop is also more efficient than using a generic flip-flop with combinational gates. In terms of area, power and delay, combinational flip-flops are desirable but they increase the design complexity. The cell library has to manage many different types of sequential elements and sometimes logic synthesis systems have difficulties in synthesizing circuits with combinational flip-flops. In a full custom design style where the complexity issue is manageable, combinational flip-flops should be used.

4.4.4 Double Edge Triggered Flip-flop

Most flip-flops in used today are called *single edge triggered flip-flop* (SETFF) because the data are loaded at only one clock edge, either rising or falling. A flip-flop can also be designed so that it loads data at both rising and falling clock edges [4.16]. Figure 4.16(b) shows such a flip-flop. Compared to a SETFF, the *double edge triggered flip-flop* (DETFF) requires slightly more transistors to implement. The flip-flop retains its data when the clock signal is not toggling.

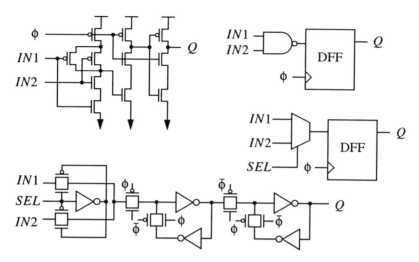

FIGURE 4.15 Combinational flip-flops.

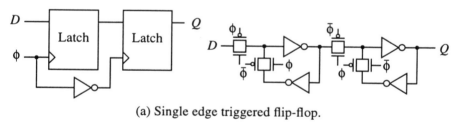

(a) Single edge triggered flip-flop.

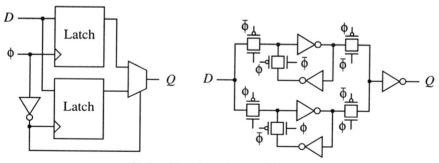

(b) Double edge triggered flip-flop.

FIGURE 4.16 Single and double edge triggered flip-flops.

Since the DETFF responds to both clock edges, the clock frequency can be halved to achieve the same computation throughput compared to SETFF. This results in overall system power reduction because clock distribution is a major source of power consumption in a synchronous computation system.

Let's analyze the power consumption of SETFF and DETFF systems. We assume that the clock network capacitance C_{clk} is identical in both cases. This is reasonable because both flip-flops typically have internal clock buffering, as shown in Figure 4.16. The power consumption of a flip-flop in the SETFF system is

$$P_s = E_{sc}f_c + E_{sd}f_d \tag{4.8}$$

where E_{sc} is the energy consumed due to clocking and E_{sd} is the additional energy consumed when the flip-flop is loaded with a different value, i.e., its output switches. The variable f_c is the clock frequency and f_d is the average frequency of the flip-flop's output. For the DETFF system, the power dissipation per flip-flop is

$$P_d = 0.5E_{dc}f_c + E_{dd}f_d \tag{4.9}$$

where E_{dc} and E_{dd} is the clock energy and load energy, respectively, of the DETFF. The frequency of the DETFF system is halved to achieve the same computation. Typically, the data rate f_d is a small fraction of the operating frequency f_c. Since the DETFF circuit is larger, the DETFF energy E_{dc}, E_{dd} is generally larger than SETFF energy E_{sc}, E_{sd}. Assuming $E_{sc} = E_{sd}$, $E_{dc} = 1.3E_{sc}$, $E_{dd} = 1.3E_{sd}$, $f_d = 0.4f_c$, we have

$$P_s = 1.4E_{sc}f_c$$
$$P_d = 1.17E_{sc}f_c = 0.84P_s \tag{4.10}$$

The above analysis only considers the power dissipation of the flip-flops. Since the clock signal of the DETFF is running at half the frequency, there is also a 2X power savings in the clock distribution network outside the flip-flops. It has been reported that the power dissipation of the DETFF is 45% less compared to a SETFF [4.17].

The obvious disadvantage of the DETFF system is the area penalty because the DETFF is about 15% larger in transistor count. The maximum operating frequency of the DETFF is less than that of the SETFF. However, since the DETFF runs at half the clock rate, the maximum effective throughput of both systems should be approximately the same. The DETFF system requires a strict 50% duty cycle clock for performance critical applications. Some logic synthesis system may also have difficulties in synthesizing DETFF circuits. The E_{sc}/E_{dc} ratio is also an important factor in applying this design style. A larger ratio favors DETFF systems because the flip-flop output frequency f_d is the same for both types of flip-flops, given the same computation. In Figure 4.16, the SETFF and DETFF have 8 and 12 transistors to be clocked, respectively. Some SETFF can be designed with significantly lower clock loading (for example, the single phase flip-flop in Figure 4.11) and thus diminish the power advantage of the DETFF system.

4.5 Low Power Digital Cell Library

Over the years, the major VLSI design focus has shifted from masks, to transistors, to gates and to register transfer level. Such upward propagation of design abstraction allows a designer to manage increasingly larger designs. It is becoming cost prohibitive to design digital circuits manually at the gate-level because of the complexity of the chip. Most digital circuits are now synthesized from some form of higher-level description automatically. The basic building blocks of gate-level synthesis are the gates or cells. Undoubtedly, the quality of the gate-level circuit synthesized depends on the quality of the cell library. In this section, we will discuss some properties of a cell library that forms the basis for good low power design.

4.5.1 Cell Sizes and Spacing

In the top-down cell-based design methodology, the trade-off among power, area and delay is performed by selecting the appropriate sizes of the cells. Sizing at the transistor level is seldom applied because of the difficulties in automated physical design. Therefore, the most important attribute that constitutes a good low power cell library is the availability of wide ranges of cell sizes for commonly used gates. As discussed in Section 4.1.2, the smallest gate that satisfies the delay requirement should be used to achieve the best power efficiency of the overall design. Thus, the *granularity* of the available gate sizes is crucial. For example, if the timing constraint commands the use of a 3X gate and the closest available gate is at 4X, the resultant choice would be 1X larger than necessary, thus wasting some power.

In the past when library cells are not designed with the power dissipation consideration, the available sizes are limited. In [4.18], up to 26 different gate sizes are used in the cell library to improve power efficiency, about four times more than that available in the traditional cell libraries. The *spacing* of the cell sizes should also be carefully chosen. The *capacitance distribution profile* of the circuit should be considered in drive strength spacing. For example, if most net capacitance is within [0.1pF, 0.5pF], there should be more cell sizes available to drive capacitances at this range. Usually, the capacitances are concentrated in the low ranges. Thus, the lower drive cells should be spaced closer than the higher drive cells, e.g., 1X, 2X, 3X, 5X, 8X, 12X, so that the size mismatches are reduced.

Today, the speed of logic synthesis is still a concern and having too many cells slows down the logic synthesis process. One way to reduce library cell count without too much compromise in quality is to have more size selections for gates that are commonly used than those that are less likely used. Another method is to use a subset of the cells for the initial synthesis runs to favor synthesis speed, followed by a full set of library cells at the final run to obtain the best circuit quality.

4.5.2 Varieties of Boolean Functions

How many different Boolean functions of n-inputs exist? The answer can be readily obtained from the truth table of an n-input Boolean function. There are 2^n entries at the output column of the truth table, which can be assigned with zeros or ones. Each zero-one assignment of the output column results in a unique Boolean function. Therefore, there are

$$M = 2^{2^n} \tag{4.11}$$

distinct Boolean functions with n-inputs. The number M is enormous even for very small values of n. For $n \geq 3$, only a small number of the M Boolean functions are available as gates in a cell library. Typically, the Boolean operations that correspond to human intuition are implemented, such as AND, OR, EXOR, AOI, OAI, etc.

The lack of varieties of Boolean functions in a cell library can result in inferior circuits to be generated. For example, if the Boolean function $Y = A\overline{B}$ were to be implemented and the *inverted-input* cells are not available, the logic synthesis system has to use an INVERTER and an AND gate to implement the function. This is shown in Figure 4.17. The two-cell implementation is less power efficient compared to the single-cell implementation because of the external routing capacitance at the output of the inverter.

FIGURE 4.17 Inverted input cells for low power cell library.

Implementing all M Boolean functions in a cell library is obviously impossible. How many functions should be included for the library to be sufficiently rich? The problem is difficult to answer and there has been a lack of theoretical analysis in this area. Among the M functions, some are degenerate in the sense that the output does not depend on all input variables. Among the non-degenerate functions, most are considered similar because of symmetry and other properties. For example, the functions $Y = A\overline{B}$ and $Y = \overline{A}B$ are identical under symmetrical equivalence and there is no need to implement both.

Some results from the studies of Boolean function classification [4.19] [4.20] may give some cues. Two Boolean functions are equivalent under permutation (called *P-equivalent*) if one function can be obtained from the other by permuting the input variables. For example, the function $Y = AB + C$ is P-equivalent to $Y = AC + B$ because we can interchange the inputs B and C of one function to obtain the other. The P-equivalence relation partitions the M Boolean functions into equivalence classes, such that the functions that belong to the same class are mutually P-equivalent. To reduce the cell count, the cell library designer only needs to implement one function from each P-equivalence class.

It is also interesting to examine the equivalent function under *input negation* (called *input negation equivalent*). The function $Y = AB$ is input negation equivalent to $Y = \overline{A}B$ and $Y = \overline{A}\,\overline{B}$ because we can invert the input variables to obtain the other functions. Similarly we can also define *output negation equivalent*, e.g., $Y = AB$ is output negation equivalent to $Y = \overline{AB}$ because one is obtained from the other by negating the output.

The equivalence relation formed by either input negation or permutation is called *NP-equivalent*. Two functions are NP-equivalent if one can be obtained from the other by any combination of input negations or variable permutations. Obviously, the number of distinct NP-equivalence classes is less than that of P-equivalence classes because P-equivalence implies NP-equivalence. Similarly, we define *NPN-equivalence* to be the equivalence classes formed by input and/or output negation and/or variable permutation. Table 4.1 shows the number of classes of n-input Boolean function under the various equivalence relation defined above. The first column is number of input variables n. The second column shows the number of possible distinct functions of n variables, i.e., $M = 2^{2^n}$ as discussed above. The third column gives the number of n-variable functions that are non-degenerate, i.e., all n variables are required to determine its output value. The fourth, fifth and sixth columns give the number of distinct functions under P-equivalence, NP-equivalence and NPN-equivalence, respectively, among the non-degenerate functions.

Even with small n, the number of distinct cells to be implemented under the various equivalence classes is large. Intuitively, a cell library that covers more classes should produce circuits with better quality because it offers more choices of distinct Boolean

TABLE 4.1 Classes of Boolean function under P, NP, and NPN-equivalence.

Input var. (n)	Num. func. 2^{2^n}	Strict n variables	P-equiv	NP-equiv	NPN-equiv
1	4	2	2	1	1
2	16	10	8	3	2
3	256	218	68	16	10
4	65536	64594	3904	380	208
5	$\approx 4.3 \times 10^9$	$\approx 4.3 \times 10^9$	$\approx 3.7 \times 10^7$	1,227,756	615,904

functions. However, there has been a lack of experimental results to support this conjecture. As an example, the ten types of three-input functions under the NPN-equivalence relation are shown in Figure 4.17.

4.6 Adjustable Device Threshold Voltage

We have seen that reducing operating voltage is the most important means for reducing power consumption of a chip due to the quadratic effect of $P = CV^2f$ equation. A first order analysis shows that as we decrease operating voltage, the gate delay t_d and the performance f_{max} (maximum clock frequency) of a chip degrade with the following equations [4.21] [4.22]

$$t_d \propto \frac{V_{dd}}{(V_{dd} - V_t)^2} \qquad (4.12)$$

$$f_{max} \propto \frac{1}{t_d} \propto \left(1 - \frac{V_t}{V_{dd}}\right)(V_{dd} - V_t) \qquad (4.13)$$

As V_{dd} gets closer to the sum of the P and N device threshold voltages V_t, the maximum operating frequency f_{max} degrades rapidly and becomes a limiting factor. Looking at the equation, an easy solution is to scale down the threshold voltage V_t proportionally to regain speed. In low voltage devices, V_t is decreased but not at the same proportion as V_{dd}. There are at least two reasons that V_t cannot be decreased proportionally. The most important reason is that as the threshold voltage decreases, the sub-threshold leakage current of the device increases exponentially (see Section 1.4.2, Figure 1.9). The second problem is that the process variation of the threshold

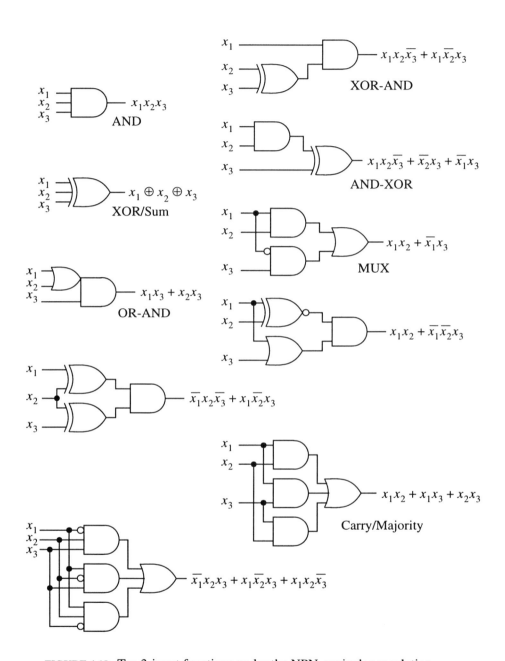

FIGURE 4.18 Ten 3-input functions under the NPN-equivalence relation.

voltage does not scale accordingly. For example, it has been shown that a ±10% threshold voltage variation tolerance at $V_{dd} = 3.3V$ translates to ±1.5% at $V_{dd} = 0.9V$ [4.23]. Such difficulties in device fabrication prevent further lowering of V_t. At operating voltage close to 1V, the threshold voltage is typically set to 0.3-0.5V to maintain an acceptable speed and leakage.

One solution to the problems is to introduce devices with different threshold voltages on the same chip. A small number of devices that are speed-critical operate at low V_t. Other devices operate at higher V_t to avoid excessive leakage current. Multiple V_t devices can be implemented by process or circuit techniques. With the process technique, the designer identifies the speed critical devices and an additional mask layer is used to mark the device channels [4.24]. By adjusting the channel doping of the marked devices, their threshold voltages can be decreased. The advantage of this is the simplicity of circuit design and that the threshold voltage of each individual device can be programmed. The disadvantage is that only a small number of threshold voltages are allowed because additional masks increase fabrication cost and complexity.

Another method to adjust V_t is by controlling the bulk bias voltage, exploiting the body effect of the MOS transistor. This does not require changes in the process but the stability of V_t may not be as good under temperature variation. The bias voltage generation circuit itself may consume a significant amount of static current. In [4.25], an adaptive mechanism is used to adjust the bias voltage generation circuit to improve V_t stability. The bias voltage needs to be routed to the device substrate. Since the bias voltage is different from the supply voltage or ground, there is a routing penalty in supplying the bias voltage. Therefore, adjusting the threshold voltage of an individual device selectively is not advisable because of the excessive routing overhead of the bias voltage. Typically, a large group of devices that are physically closed together, such as devices in a pad cell or a functional module, is adjusted.

REFERENCES

[4.1]   J. Fishburn and A. Dunlop, "TILOS: A Polynomial Programming Approach to Transistor Sizing," *Proceedings of International Conference on Computer-Aided Design*, pp. 326-328, 1985.

[4.2]   C. Mead and L. Conway, *Introduction to VLSI Systems*, Addison-Wesley, 1980.

[4.3]   N. Weste and K. Eshraghian, *Principles of CMOS VLSI Design: A System Perspective*, Second Edition, Addison-Wesley, 1993.

[4.4]   D. Zhou and X. Liu, "On the Optimal Drivers of High-Speed Low Power IC's," *International Journal of High Speed Electronics and Systems*, vol. 7, no. 2, pp. 287-303, Jun. 1996.

[4.5]   M. Borah, R. Owens and M. Irwin, "Transistor Sizing for Low Power CMOS Circuits," *IEEE Transactions on Computer-Aided Design of Integrated Circuits and Systems*, vol. 15, no. 6, pp. 665-671, Jun. 1996.

[4.6]   D. Chen and M. Sarrafzadeh, "An Exact Algorithm for Low Power Library-Specific Gate Resizing," *Proceedings of Design Automation Conference*, pp. 783-788, 1997.

[4.7]   P. Gronowski, W. Bowhill, *et al.*, "A 433-Mhz 64-b Quad-Issue RISC Microprocessor," *IEEE Journal of Solid-State Circuits*, vol. 31, no. 11, pp. 1687-1696, Nov. 1996.

[4.8]   S. Prasad and K. Roy, "Circuit Optimization for Minimization of Power Consumption under Delay Constraint," *Digest of Technical Papers, IEEE Symposium on Low Power Electronics*, pp. 15-20, 1994.

[4.9]   A. Glebov, D. Blaauw and L. Jones, "Transistor Reordering for Low Power CMOS Gates using an SP-BDD Representation," *Proceedings of International Symposium on Low Power Design*, pp. 161-166, 1995.

[4.10] M. Tachibana, *et al.*, "Power and Area Optimization by Reorganizing CMOS Complex Gate Circuits," *Proceedings of International Symposium on Low Power Design*, pp. 155-160, 1995.

[4.11] D. Dobberpuhl, *et al.*, "A 200-MHz 64-b Dual-Issue CMOS Microprocessor," *IEEE Journal of Solid-State Circuits*, vol. 27, no. 11, pp. 1555-1566, Nov. 1992.

[4.12] J. Yuan and C. Svenson, "High-speed CMOS circuit techniques," *IEEE Journal of Solid-State Circuits*, vol. 24, no. 1, pp. 62-70, Feb. 1989.

[4.13] U. Ko and P. Balsara, "High Performance Energy Efficient Master-Slave Flip-Flop Circuits," *Digest of Technical Papers, IEEE Symposium on Low Power Electronics*, pp. 16-17, 1995.

[4.14] J. Montanaro, R. Witek, *et al.*, "A 160-MHz 32-b 0.5W CMOS RISC microprocessor," *IEEE Journal of Solid-State Circuits*, vol. 31, no. 11, pp. 1703-1714, Nov. 1996.

[4.15] A. Reyes, S. Millman and S. Tyler, "Low Power Flip-Flop Circuit and Method Thereof," United States Patent, no. 5,498,988, Mar. 1996.

[4.16] R. Hossain, L. Wronski and A. Albicki, "Low Power Design Using Double Edge Triggered Flip-Flop," *IEEE Transactions on VLSI Systems*, vol. 2, no. 2, pp. 261-265, June 1994.

[4.17] R. Llopis and M. Sachdev, "Low Power Testable Dual Edge Triggered Flip-Flops," *Digest of Technical Papers, International Symposium on Low Power Electronics and Design*, pp. 341-345, 1996.

[4.18] A. Correale, "Overview of the Power Minimization Techniques Employed in the IBM PowerPC 4xx Embedded Controllers," *Proceedings of International Symposium on Low Power Design*, pp. 75-80, 1995

[4.19] S. Hurst, D. Miller and J. Muzio, *Spectral Techniques in Digital Logic*, Academic Press, London, 1985.

[4.20] S. Hurst, *The Logical Processing of Digital Signals*, Crane, Russak and Company, New York, 1978.

[4.21] A. Chandrakasan, S. Sheng and R. Brodersen, "Low-power CMOS Digital Design," *IEEE Journal of Solid-State Circuits*, vol. 27, no. 4, pp. 473-484, Apr. 1992.

[4.22] M. Horowitz, T. Indermaur and R. Gonzalez, "Low-Power Digital Design," *Digest of Technical Papers, IEEE Symposium on Low Power Electronics*, pp. 8-11, 1994.

[4.23] S. Sun and P. Tsui, "Limitation of CMOS Supply Voltage Scaling by MOSFET Threshold Voltage Variation," *Proceedings of IEEE Custom Integrated Circuits Conference*, pp. 267-270, 1994.

[4.24] J. Ma, *et al.*, "A Graded-channel MOS (GCMOS) VLSI Technology for Low Power DSP Applications," *Digest of Technical Papers, International Symposium on Low Power Electronics and Design*, pp. 129-132, 1996.

[4.25]   V. von Kaenel, *et al.*, "Automatic Adjustment of Threshold and Supply Voltages for Minimum Power Consumption in CMOS Digital Circuits," *Digest of Technical Papers, IEEE Symposium on Low Power Electronics*, pp. 78-79, 1994.

CHAPTER 5   *Logic*

Logic design was once the primary abstraction level where automatic design synthesis begins. With the advance of logic synthesis tools and structured VLSI design practice today, logic design is seldom performed manually. However, logic analysis is still the basis of VLSI chip design. A strong foundation in logic design is the key to produce high quality chips.

Even with the use of text-based *hardware description language* design style, there are still many techniques for the VLSI designer to reduce power at the logic level. The most prevalent theme in logic level power optimization techniques is the reduction of switching activities. Switching activities directly contribute to the charging and discharging capacitance, and the short-circuit power. The choice of logic encoding, data representation and Boolean function implementation has significant effects on the power dissipation of a digital circuit. Some switching activities are the result of unspecified or undefined behavior of a logic system that are not related to its proper operation. Such stray switching activities should be eliminated or reduced if possible. However, suppressing unnecessary activities usually requires additional hardware logic that increases the area and consumes power. The challenge is to justify the low power techniques via intelligent analysis and trade-off while managing the increased design complexity.

5.1 Gate Reorganization

In Section 4.3, we discussed the techniques for transistor network restructuring and reorganization for power efficiency. The reorganization idea is not limited to transistor networks only since the same problem exists in gate-level networks. In this case, we have a gate-level network implemented with a set of logic gates. Network reorganization is applied to the gate-level network to produce logically equivalent networks with different qualities for power, area and delay. A related problem known as *technology mapping* is very similar to gate reorganization except that the original network is expressed in a generic form such as two-input NAND gates only. The reorganized network hopefully has better power efficiency than the original network.

The complexity of the gate-reorganization problem limits manual solution to small circuits only. Nevertheless, we will introduce several simple logic restructuring techniques so that the reader can have an appreciation of the problem. Most gate reorganization tasks today are performed by automated software in the logic synthesis system. Gate-level reorganization is a central operation of logic synthesis. In the past, logic synthesis systems perform optimization and trade-off primarily for area and delay only. Power optimization requirements add another dimension to the complexity of the problem. Commercial and academic logic synthesis tools that automatically perform power-area-delay optimization have been introduced [5.1] [5.2].

5.1.1 Local Restructuring

In the most general sense, gate reorganization is an operation to transform one logic circuit to another that is functionally equivalent. Since there are infinite number of ways to implement a logic circuit, a transformation operation that drastically changes the delay and power characteristics of the circuit has no practical value. The result of such drastic transformation is like randomly generating functionally equivalent circuits without an optimization goal. Therefore, most logic restructuring techniques use *local restructuring rules* to transform one network to another. Such local transformations only affect very few of gates and wires in the circuit. Typically these affected gates are directly connected in the logic network. The incremental changes in power, delay and area of such local transformation can be evaluated easily.

From the hardware description language, a logic synthesis system typically constructs an initial logic network that is believed to be quite good. Then, the gate reorganization procedure applies a series of local transformations to the network. Many networks are generated and evaluated and the best circuit is retained. Some basic transformation operators are:

1. Combine several gates into a single gate.
2. Decompose a single gate into several gates.
3. Duplicate a gate and redistribute its output connections.
4. Delete a wire.
5. Add a wire.
6. Eliminate unconnected gates.

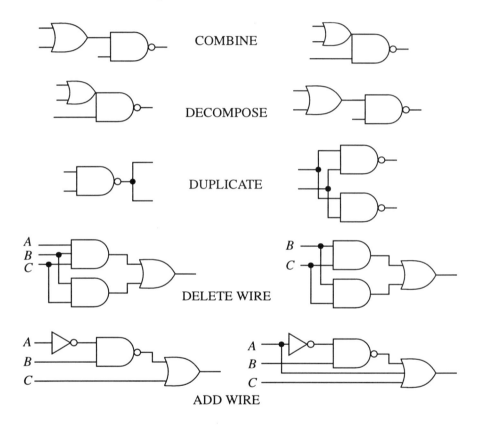

COMBINE

DECOMPOSE

DUPLICATE

DELETE WIRE

ADD WIRE

FIGURE 5.1 Local transformation operators for gate reorganization.

Figure 5.1 illustrates some local transformation operators. It can be easily verified that the circuits on the left and those on the right are logically identical. Some of the transformation operators have a direct contribution to power dissipation or propagation delay. The COMBINE operator can be used to "hide" high-frequency nodes inside the cell so that the node capacitance is not being switched. The DECOMPOSE and DUPLICATE operators help to separate the critical path from the non-critical ones so that the latter can be sized down. The DELETE WIRE operator reduces the circuit size and the ADD WIRE operator helps to provide an intermediate circuit that may eventually lead to a better one.

Many other equivalent transformation operators have been proposed [5.3] [5.4] [5.5] [5.6] [5.7]. Repeated applications of the local transformation operators will generate gate-level networks of different timing, area and power trade-off. In an automated system, some cost functions are applied to evaluate the qualities of the networks and the best one is retained as the final result.

5.2 Signal Gating

Signal gating refers to a class of general techniques to mask unwanted switching activities from propagating forward, causing unnecessary power dissipation. Most signal gating techniques are applied at the logic level because switching activities of the signals can be easily analyzed. The probabilistic techniques developed in Chapter 3 are often used for switching activity analysis.

There are many different ways to implement signal gating. The simplest method is to put an AND/OR gate at the signal path to stop the propagation of the signal when it needs to be masked. Another method is to use a latch or flip-flop to block the propagation of the signal. Sometimes, a transmission gate or a tristate buffer can be used in place of a latch if charge leakage is not a concern. Figure 5.2 shows the various logic implementations of signal gating. The signals at the bottom of the circuits are control signals used to suppress the source signal on the left from propagating to the gated signal on the right.

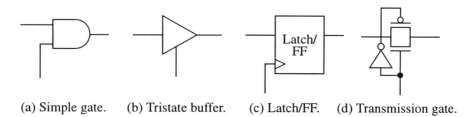

(a) Simple gate.   (b) Tristate buffer.   (c) Latch/FF.   (d) Transmission gate.

FIGURE 5.2 Various logic implementations of signal gating.

All signal gating methods require control signals to stop the propagation of switching activities, as indicated by the lower terminals in Figure 5.2. The generation of control signals requires additional logic circuitry at the controllers. Thus, the additional power required to generate the control signals needs to be considered to see if the gating leads to overall power saving. In general, if the relative enable/disable frequency of the control signal is much lower than the frequency of the gated signal, this technique can save power because many signal switching activities are blocked with only a few additional activities at the control signal. If a group of signals can share a common control, such as on a bus or a clock tree network, the power saving will be remarkable.

The identification of signals to be gated is highly application dependent. Major considerations are the power saving anticipated and the complexity of the control signal generation circuit. Good candidates for signal gating are clock signals, address or data busses and signals with high frequency or glitches. Specific techniques for clock and bus gating will be discussed in later chapters.

5.3 Logic Encoding

The logic designer of a digital circuit often has the freedom of choosing a different encoding scheme as long as the functional specification of the circuit is met. For example, an 8-bit counter can be implemented using the Binary counting sequence or the Gray code sequence. Different encoding implementations often lead to different power, area and delay trade-off. Usually, encoding techniques require the knowledge of signal statistics in order to make design decisions. This section discusses some techniques for using different logic encoding to achieve low power consumption.

5.3.1 Binary versus Gray Code Counting

Consider two n-bit counters implemented with Binary and Gray code counting sequences. The counting sequences of the two counters are listed in Table 5.1. In a full counting cycle, the number of transitions of a Binary counter is

$$B_n = 2 (2^n - 1) \qquad (5.1)$$

TABLE 5.1 Binary and Gary code counting sequences.

Binary code		Gray code	
Sequence	No. toggles	Sequence	No. toggles
0 0 0	3	0 0 0	1
0 0 1	1	0 0 1	1
0 1 0	2	0 1 1	1
0 1 1	1	0 1 0	1
1 0 0	3	1 1 0	1
1 0 1	1	1 1 1	1
1 1 0	2	1 0 1	1
1 1 1	1	1 0 0	1

TABLE 5.2 Toggle activities of Binary versus Gray code counter.

No. bits	No. of toggles			B_n / G_n
	Binary $B_n = 2(2^n - 1)$	Gray $G_n = 2^n$		
1	2	2		1
2	6	4		1.5
3	14	8		1.75
4	30	16		1.88
5	62	32		1.94
6	126	64		1.99
∞	-	-		2.00

For example, the 2-bit Binary counting cycle 00, 01, 10, 11, back to 00, has $1 + 2 + 1 + 2 = 6$ bit-flips. However, since a Gray code sequence only has one bit-flip in each count increment, the number of transitions in a full counting sequence is

$$G_n = 2^n \tag{5.2}$$

Table 5.2 shows the ratio of B_n to G_n for different values of n. When n is large, the Binary counter has twice as many transitions as the Gray counter. Since power dissipation is related to toggling activities, a Gray counter is generally more power efficient than a Binary counter.

5.3.2 Bus Invert Encoding

Bus invert encoding [5.8] is a low power encoding technique that is suitable for a set of parallel synchronous signals, e.g., off-chip busses. The architecture of bus invert encoding is illustrated in Figure 5.3. At each clock cycle, the data sender examines the current and next values of the bus and decides whether sending the *true* or the *compliment* signal leads to fewer toggles. Since the data signals on the bus may be complemented, an additional polarity signal is sent to the bus receiver to decode the bus data properly. For example, if the current value of the bus is 0000 and the next value is 1110, it will send the complement of the next value, 0001, and assert the polarity signal. The assertion of the polarity signal tells the receiver to invert the received bus signals.

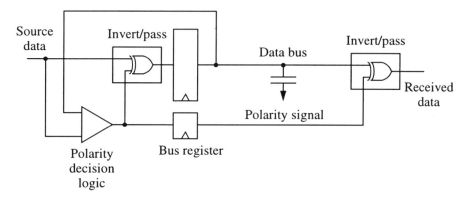

FIGURE 5.3 Architecture of bus invert encoding.

The implementation of the bus invert encoding technique is quite simple. We need a register to store the current value of the bus. A polarity decision logic module, which contains combinational logic only, compares the current and next value of the bus and decides whether to invert the bus signals. A bank of XOR gates at the sending and receiving ends inverts the bus signals.

The overhead associated with the bus invert technique is obviously the additional area, power and sometimes propagation delay to implement the polarity decision logic, the invert/pass gates and the polarity signal. If the capacitance of the bus is large enough, the power saving gained from reduced bus signal switching will offset the additional power dissipation required by the polarity and invert logic.

Let us analyze the efficiency of the bus invert encoding scheme by examining the reduction in signal toggle activities on the bus. Compared to a regular bus, the maximum number of toggling bits of the inverted bus is reduced from n to $n/2$. The reasoning is as follows: if there were more than $n/2$ toggling bits in the inverted bus, we would have inverted the data signals, thus reducing the toggling bits below $n/2$. The power efficiency of the inverted bus depends on the signal statistics of bus data. Under the uniform random signal condition, the toggle reduction of the bus invert scheme has an upper bound of 25%, as we will show below.

We use the memoryless random signal model described in Section 3.2.1 for analysis. We assume that each bit of the bus has the uniform random probability distribution and is uncorrelated in time. This means that the current value of the bus is independent of its previous values and all bits are mutually uncorrelated. Under this assumption, the probability of a k-bit transition on an n-bit regular bus is

$$P_k = \frac{1}{2^n}\binom{n}{k} \tag{5.3}$$

The expected number of transitions $E[P]$ of the regular bus is thus

$$E[P] = \sum_{k=0}^{n} k P_k = \frac{n}{2} \tag{5.4}$$

Using the bus invert scheme, we introduce one polarity bit to the bus. We further assume that n is even. (If n were odd, the analysis is similar.) For the $(n+1)$-bit inverted bus, the number of bit transitions at any given clock cycle is never more than $n/2$. At each clock cycle, if there is a k-bit transition on the inverted bus, one of the following two conditions must occur:

1. The polarity bit does not toggle: the probability of this condition is identical to that of a k-bit transition on a regular bus P_k.
2. The polarity bit toggles: this means that there are $k-1$ bit transitions in the inverted bus, which implies that there are $n-k+1$ bit transitions in the corresponding regular bus. This probability is given by $P_{(n-k+1)} = P_{(k-1)}$.

Thus, the probability of a k-bit transition on the $(n+1)$-bit inverted bus is

$$Q_k = P_k + P_{(n-k+1)} = P_k + P_{(k-1)} = \frac{1}{2^n}\binom{n+1}{k} \tag{5.5}$$

The expected number of transitions $E[Q]$ on an inverted bus is thus

$$E[Q] = \sum_{k=0}^{n/2} k\, Q_k \tag{5.6}$$

Table 5.3 lists the values of $E[P]$, $E[Q]$ and the ratio $E[Q]/E[P]$ for the same value of n. It shows that the ratio increases with the size of the bus. For example a 16-bit inverted bus only reduces the toggle activity by 15% compared to the regular bus. The best efficiency is achieved when $n = 2$. This seems to be counter intuitive but true. The reason is that the probability of asserting the invert signal is less on a wider bus, thus leading to less saving. This is a direct result of our uniform probability assumption. However, if the bus has a high tendency of making more than $n/2$ bit transitions, the toggle ratio may improve.

TABLE 5.3 **Efficiency of bus invert encoding under uniform random signal.**

Num. bits	Regular bus $E[P]$	Invert bus $E[Q]$	Invert / Regular $E[Q]/E[P]$
2	1	0.75	0.75
4	2	1.56	0.781
8	4	3.27	0.817
16	8	6.83	0.854
32	16	14.19	0.886
64	32	29.27	0.915
128	64	59.96	0.937
256	128	122.1	0.954
∞	-	-	1.00

The design decision to apply bus invert technique is dependent on signal statistics and the overhead associated with the polarity decision logic, the polarity signal and the invert/pass gates. If the design uses logic synthesis, the bus invert scheme can be easily implemented and compared to the regular bus to decide which design is better. A gate-level power analysis tool can be used to verify if the inverted bussing scheme leads to acceptable power reduction. In general, bus invert encoding adds a substantial amount of circuitry to the final implementation. Unless the bus capacitance is large enough, the overhead may not be acceptable. This technique is particularly suited for off-chip busses, because of large loading capacitance and small overhead in additional on-chip logic. Also, since the maximum number of transitions on the bus is reduced from n to $n/2$, this technique is suited when the peak current needs to be reduced.

5.4 State Machine Encoding

A state machine is an abstract computation model that can be readily implemented using Boolean logic and flip-flops, as shown in Figure 5.4. In today's logic synthesis environment, a *state transition graph* is specified by the designer and the synthesis system will produce a gate-level circuit based on the machine's specification. Sometimes, the behavior of a state machine is described in a high-level description language suitable for synthesis, in which case the description is either generated manually or obtained from a state machine compiler.

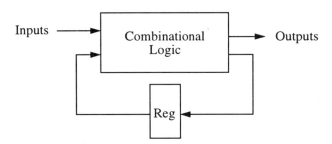

FIGURE 5.4 Hardware architecture of a synchronous state machine.

The state transition graph is a functional description of a machine specifying the inputs and outputs of the machine under a particular state and its transition to the next state. The very first step of a state machine synthesis process is to allocate the state register and assign binary codes to represent the symbolic states. This process is called the *encoding* of a state machine. The encoding of a state machine is one of the most important factors that determine the quality (area, power, speed, etc.) of the gate-level circuit [5.9] [5.10] [5.11] [5.12] [5.13] [5.14] [5.15].

During state machine implementation, the only optimization decision regarding the state register is the number of bits to be assigned. Often, the choice is to use the minimum number of bits that are enough to represent all states or one slightly greater than that [5.14]. Thus, the real optimization issue that determines the quality of a state machine is the synthesis of the combinational logic network. When a state transition graph is properly encoded, the state machine synthesis problem becomes a combinational logic synthesis problem because the Boolean functions of the combinational network have been completely specified. Thus, the encoding of a state machine has a profound impact on the quality of the final implementation.

5.4.1 Transition Analysis of State Encoding

Several key parameters have been observed to be very important to the power efficiency of state encoding. One such parameter is the expected number of bit transitions in the state register. Another parameter is the expected number of transitions of output signals.

Consider two functionally identical state machines $M1$ and $M2$ with different encoding shown in Figure 5.5. The binary codes in the state bubbles indicate the state encoding. The labels at the state transition edges represent the probabilities that transitions will occur at any given clock cycle. The sum of all edge probabilities equals to unity. The expected number of state-bit transitions $E[M]$ is given by the sum of products of edge probabilities and their associated number of bit-flips as dictated by the encoding. For example, the machine $M1$ has two edges with two bit transitions (00 and 11) at probabilities 0.3 and 0.4. The contribution of the edges to the expected transition is $2(0.3 + 0.4)$. The expected transitions per clock cycle for the two machines are given in the figure.

In general, machines with lower $E[M]$ are more power efficient because:

1. fewer transitions of the state register lead to lower power dissipation, and

2. fewer transitions are propagated into the combinational logic of the machine.

However, a state encoding with the lowest $E[M]$ may not be the one that results in the lowest overall power dissipation. The reason is that the particular encoding may require more gates in the combinational logic, resulting in more signal transitions and power.

Figure 5.6 shows the synthesized area and power dissipation of some randomly encoded state machines [5.15]. All machines are functionally equivalent and meet timing constraints except that their state encoding is different. Each point in the figure represents a state machine synthesized with a particular encoding, resulting in a particular area and power dissipation. Notice that the points seem to lie on the main diagonal. This suggests that the area and power dissipation of a state machine are correlated when the state encoding is varied. A larger state machine seems to dissi-

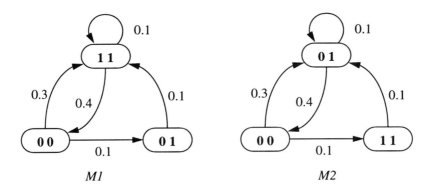

Expected state transitions:

$$E[M1] \; = \; 2\,(0.3 + 0.4) + 1\,(0.1 + 0.1) \; = \; 1.6$$
$$E[M2] \; = \; 1\,(0.3 + 0.4 + 0.1) + 2\,(0.1) \; = \; 1.0$$

FIGURE 5.5 Functionally identical state machines with different encoding.

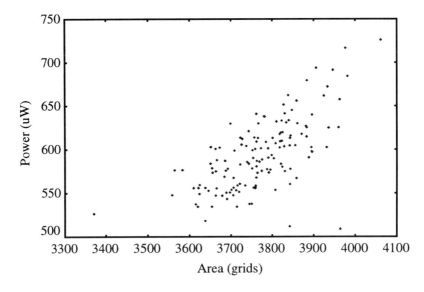

FIGURE 5.6 Effect of state encoding on synthesized area and power dissipation.

pate more power possibly because more gates and nodes are toggling in the circuit. However, within the same power dissipation level, there are some variations in area estimates.

5.4.2 Output Don't-care Encoding

In some state machine specifications, an output signal is often undefined or *don't-care* at some states and thus can have a logic 0 or 1. A proper assignment of the don't-care signal can reduce the expected transitions of the output signal and saves power. Like the state encoding problem, the assignment of the don't-care signal may present conflicting constraints to the optimization goal. The synthesis system can generally exploit the output don't-care, by fixing the signal to a logic value that leads to better area efficiency. Unfortunately, this problem has not been well studied and there are few synthesis tools that can exploit the output don't-care assignment for power efficiency.

5.4.3 Design Trade-offs in State Machine Encoding

The area-power trade-off of state machine encoding is not well understood and is heavily influenced by the logic synthesis system. State encoding affects the power dissipation as well as the area of the machine. Since area and power dissipation interacts in a complex manner, the result of a particular encoding is difficult to predict. Also, given the probabilities, the theoretical problem of finding encoding that minimizes the expected transition $E[M]$ is extremely difficult. However, in practice, near optimal encoding can often be found by applying some simple heuristic rules [5.14].

Most commercial logic synthesis system can perform automatic state encoding for the goal of area minimization. However, the encoding chosen by the synthesis system may not be desirable for power dissipation because the expected transition is high. If we encode the states to minimize the expected transition, the area is often not good because the logic synthesis system has lost the freedom of assigning state codes that favor area optimization.

One practical solution is to balance the area-power trade-off by encoding only a subset of states that spans the high probability edges. For example, if several edges constitute more than 90% of all state transitions, we can encode the states incident to the high probability edges to reduce $E[M]$. The remaining state codes can be left to the logic synthesis system so that it has the freedom of choosing the proper encoding for better area efficiency. The chosen encoding does not drastically alter $E[M]$ of the machine because they only affect the low probability transitions. In practice, most state machines tend to have this skewed edge probability characteristics where a few edges dominate the state transitions probability. As an example, the edges incident to the "reset" or "interrupt" states of a CPU controller have very low transition probabilities; while states like "instruction fetch" and "memory access" have very high probabilities. If the transition probabilities do not exhibit such skewed property, the power gained from different state encoding may not be as drastic.

It should be mentioned that, unlike area optimization, the power optimization of state and output don't-care encoding requires the knowledge of the probabilities of state transitions and input signals. Occasionally, a designer can predict such statistical quantities manually with high confidence. Otherwise, a behavior-level simulation of the state machine can be done to measure the probabilities. Without any knowledge of the statistical behavior of the machine, power optimization of state machine encoding is impossible. In some systems, state machines only consume a small portion of total power dissipation and the effect of state encoding is negligible.

5.5 Precomputation Logic

Precomputation logic optimization, first proposed in [5.16], is a method to trade area for power in a synchronous digital circuit. The principle of precomputation logic is to identify logical conditions at some inputs to a combination logic that is invariant to the output. Since those input values do not affect the output, the input transitions can be disabled to reduce switching activities.

5.5.1 Basics of Precomputation Logic

One variant of the precomputation architecture is shown in Figure 5.7. R_1 and R_2 are registers with a common clock feeding a combinational logic circuit with a known Boolean function $f(X)$. Due to the nature of the function $f(X)$, there may be some conditions under which the output of $f(X)$ is *independent* of the logic value of R_2. Under such condition, we can disable the register loading of R_2 to avoid causing unnecessary switching activities, thus conserving power. The Boolean function $f(X)$ is correctly computed because it receives all required value from R_1. To generate the load-disable signal to R_2, a precomputation Boolean function $g(X)$ is required to detect the condition at which $f(X)$ is independent of R_2. We will discuss how to identify the independence condition given an arbitrary Boolean function $f(X)$. By definition, $g(X)$ depends on the input signals of R_1 only because the load-disable condition is independent of R_2. Otherwise, $f(X)$ will depend on the inputs of R_2 when the load-disable signal is active.

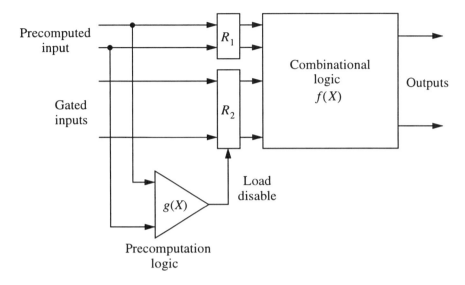

FIGURE 5.7 A variant of precomputation logic.

A remarkable example of precomputation logic is the Binary Comparator function $f(A, B)$ that computes $A > B$, as shown in Figure 5.8 [5.16]. Let inputs $A_1, ..., A_n$ and $B_1, ..., B_n$ be n-bit signals that represent binary values. The output $f(A, B)$ is logic 1 if the binary value A is greater than B. We choose R_1 to be (A_n, B_n) and R_2 to be the other signals. The signals A_n and B_n are the most significant bits of the input values. An obvious precomputation function is $g(X) = A_n \oplus B_n$. When $A_n \oplus B_n = 1$, meaning the two bits differ, the output of $f(A, B)$ can be determined without the input bits of R_2. If $A_n = 1, B_n = 0$, the output is 1 independent of the least significant bits of A or B. Again, if $A_n = 0, B_n = 1$, the output is 0 regardless of the inputs of R_2. When $A_n \oplus B_n = 0$, R_2 cannot be disabled because its signals are required to compute the output $f(A, B)$.

A simple statistical analysis may help to convince us that the precomputation logic implementation of Figure 5.8 is very attractive. Assuming uncorrelated input bits with uniform random probabilities where every bit has an equal probability of zero or one (see the memoryless random signal model in Section 3.2.1). There is a 50% probability that $A_n \oplus B_n = 1$ and the register R_2 is disabled in 50% of the clock cycles. Therefore, with only one additional 2-input XOR gate, we have reduced the signal switching activities of the $2n - 2$ least significant bits at R_2 to half of its original expected switching frequency. Also, when the load-disable signal is asserted, the combinational logic of the comparator has fewer switching activities because the out-

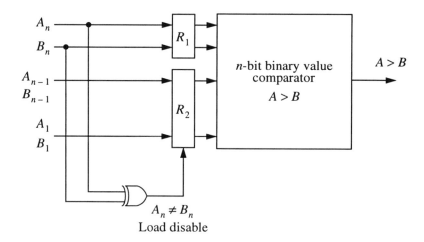

FIGURE 5.8 Binary comparator function using precomputation logic.

puts of R_2 are not switched. The extra power required to compute $A_n \oplus B_n$ is negligible compared to the power saving even for moderate size of n.

From the above discussion, it is obvious that a designer needs to have some knowledge of the input signal statistics to apply the precomputation logic technique. In Figure 5.8, if the probability of $A_n \oplus B_n$ is close to zero, the precomputation logic circuit may be inferior, in power and area, compared to direct implementation. Experimental results [5.16] have shown up to 75% power reduction with an average of 3% area overhead and 1 to 5 additional gate-delay in the worst-case delay path.

5.5.2 Precomputation Condition

Given $f(X)$, R_1 and R_2, there is a systematic method to derive a precomputation function $g(X)$. Let $f(p_1, \ldots, p_m, x_1, \ldots, x_n)$ be the Boolean function where $p_1, \ldots, p_m$ are the precomputed inputs corresponding to R_1 and $x_1, \ldots, x_n$ are the gated inputs corresponding to R_2. Let f_{x_i} ($f_{\overline{x_i}}$) be the Boolean function obtained by substituting $x_i = 1$ ($x_i = 0$) in the function f. Define

$$U_{x_i} f = f_{x_i} \cdot f_{\overline{x_i}} \tag{5.7}$$

Note that the functions f_{x_i} and $f_{\overline{x_i}}$ are new Boolean functions constructed from f and do not contain the variable x_i. The function $U_{x_i} f$ is the Boolean AND operation of the two functions f_{x_i} and $f_{\overline{x_i}}$, and by definition also does not contain the variable x_i.

The Boolean function $U_{x_i} f$ is called the *universal quantification* of f with respect to the input variable x_i. An important observation with the universal quantification function is that $U_{x_i} f = 1$ implies $f = 1$ regardless of the value of x_i. This can be observed with the help of the Shannon's decomposition of f with respect to x_i

$$f = x_i f_{x_i} + \overline{x_i} f_{\overline{x_i}} \tag{5.8}$$

The condition $U_{x_i} f = 1$ means that $f_{x_i} = f_{\overline{x_i}} = 1$ and therefore $f = 1$ regardless of the value of x_i. Let

$$g_1 = U_{x_1} U_{x_2} \dots U_{x_n} f \tag{5.9}$$

The Boolean function g_1 is constructed by first taking the universal quantification of f with x_n to obtain $U_{x_n} f$, then applying the universal quantification of $U_{x_n} f$ with x_{n-1} to obtain the second function, etc. The procedure is applied iteratively until all variables have been exhausted and we obtain the function g_1. We also note that the function g_1 does not contain variables $x_1, \dots, x_n$.

From the properties of the unification quantification operator, we see that the condition $g_1 = 1$ implies that $f = 1$ regardless of the values of $x_1, \dots, x_n$. This means that $g_1 = 1$ is one of the conditions in which the output of f is *independent* of the input values of $x_1, \dots, x_n$.

Similarly we can define

$$g_0 = U_{x_1} U_{x_2} \dots U_{x_n} \overline{f} \tag{5.10}$$

Here, g_0 is obtained by applying universal quantification to the compliment function of f. The condition $g_0 = 1$ implies $f = 0$ regardless of the values of $x_1, \dots, x_n$. Thus, the function $g = g_1 + g_0$ is a precomputation function in which $g = 1$ means that we can disable the loading of gated inputs $x_1, \dots, x_n$ into R_2 because the output f is independent of the gated inputs.

The precomputation function g only contains the precomputed variables $p_1, \dots, p_m$. In general, the function g may not exist because the universal quantification of f may evaluate to a logic 0. For example, the universal quantification of any variable of $f = x_1 \cdot x_2$ is always 0 and no precomputation function exists.

At first glance, the function g obtained from the above process is just one function among the many precomputation functions. Indeed, there may be another precomputation function g' that qualifies. However, it can also be shown that if $g = g_1 + g_0$ exists as defined by Equations (5.9) and (5.10), it maximizes the probability of load-disable condition [5.16]. In other words, the function g contains the most number of one's in its truth table among all precomputation functions qualified. Maximizing the load-disable probability is desirable because it leads to fewer activities in the combinational logic of f and the gated register R_2. However, it is difficult to predict if a function that maximizes the load-disable probability is also one that minimizes the total power consumption. This is because the power dissipation of the precomputation logic g itself cannot be ignored. An unoptimized precomputation function g' may require fewer logic gates so that it could lead to better total power consumption. In practice, it makes sense to explore other precomputation functions for better total power dissipation.

5.5.3 Alternate Precomputation Architectures

Many precomputation schemes have been proposed in [5.16]. One scheme based on the Shannon's decomposition, is shown in Figure 5.9. The Shannon's decomposition states that a Boolean function $f(x_1, ..., x_n)$ can be decomposed with respect to the variable x_i as follows

$$f(x_1, ..., x_n) = x_i f_{x_i} + \overline{x_i} f_{\overline{x_i}} \tag{5.11}$$

The equation allows us to use x_i as the load-disable signal as shown in Figure 5.9. When $x_i = 0$ ($x_i = 1$), the inputs to the logic block f_{x_i} ($f_{\overline{x_i}}$) can be disabled. The multiplexor selects the output of the combinational logic block that is active. This means that only one combinational logic block is activated at any clock cycle. Power saving is achieved if each of the two decomposed logic blocks consumes less power than a direct implementation of $f(x_1, ..., x_n)$. However, the precomputation architecture consumes more area and delay in general.

Another precomputation architecture for combinational logic is shown in Figure 5.10. The original logic implementation is shown in Figure 5.10(a) and the low power implementation is shown in (b). This architecture is also called *guarded evaluation* in [5.17] because some inputs to the logic block C_2 are isolated when the signals are not required, to avoid unnecessary transition. Transmission gates may be used in place of the latches if the charge storage and noise immunity conditions permit.

5.5.4 Design Issues in Precomputation Logic Technique

The basic design steps with precomputation logic are as follows:

1. Select precomputation architecture.

2. Determine the precomputed inputs R_1 and gated inputs R_2 given the function $f(X)$.

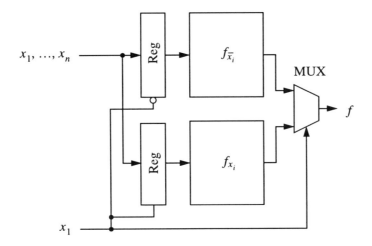

FIGURE 5.9 A precomputation architecture based on Shannon's decomposition.

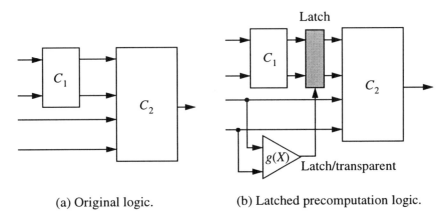

(a) Original logic.   (b) Latched precomputation logic.

FIGURE 5.10 Latch-based precomputation architecture.

3. With R_1 and R_2 selected, find a precomputation logic function $g(X)$. Note that $g(X)$ is not unique and the choice greatly affects the power efficiency. The function $g(X)$ may also fail to exist for poor choices of R_1 and R_2.

4. Evaluate the probability of precomputation condition and the potential power savings. Make sure that the final circuit is not overwhelmed by the additional logic circuitry and power consumption required to compute $g(X)$.

After R_1, R_2 and $g(X)$ are determined, the precomputation logic can be synthesized using a logic synthesis tool. The partition of input variables to precomputed and gated input sets is not trivial. To determine the best partition of input variables R_1 and R_2, an exhaustive enumeration and approximation method has been proposed [5.16], in which the objective is to maximize the probability of $g(X) = 1$.

We have presented the precomputation technique using only single output Boolean functions. For multiple output Boolean functions, the technique can be easily generalized by treating each output signal j independently and derive the bit precomputation function $g_j(X)$. The actual precomputation function $g(X)$ is the product of $g_j(X)$ over all j. However, in the case of multiple output function, the probability of precomputation condition is lowered because all output bits must agree to be precomputed. The applicability of precomputation function to multiple output Boolean function has not been well studied.

In practice, the selection of R_1, R_2 and precomputation logic $g(X)$ depends heavily on the designer's instinct based on his knowledge of the input signal statistics. If the input signal statistics is unknown, the precomputation method cannot be applied because it is impossible to evaluate the power savings resulted from gating some inputs. Another prerequisite in using precomputation logic is the availability of CAD tools for area and power estimation. The power saved from input gating and the additional power dissipated in precomputation logic have to be assessed carefully so that the area overhead can be justified.

Typically, the number of precomputed inputs R_1 should be much smaller than the gated inputs R_2 for gross power saving. Power reduction is generally more dramatic when the Boolean functions have a small number of outputs and deep logic implementation, i.e., the chains of logic gates from the inputs to outputs are long. To date, there is no commercial synthesis tool that automates the precomputation logic design process.

Precomputation is by definition creating redundant logic. The logic circuits that perform precomputation generally pose difficulties in testing. For example, in Figure 5.7, the load-disable signal is not observable unless we implement a full scan of the flip-flops. This may limit its wide spread usage. Besides testability, timing may also become a problem because the precomputation logic might be on a critical delay path of the circuit, limiting its performance.

REFERENCES

[5.1] S. Iman and M. Pedram, "POSE: Power Optimization and Synthesis Environment," *Proceedings of Design Automation Conference*, pp. 21-26, 1996.

[5.2] S. Krishnamoorthy and A. Khouja, "Efficient Power Analysis of Combinational Circuits," *Proceedings of Custom Integrated Circuit Conference*, pp. 393-396, 1996.

[5.3] D. Pradhan, M. Chatterjee, M. Swarna and W. Kunz, "Gate-Level Synthesis for Low-Power using New Transformations," *Digest of Technical Papers, International Symposium on Low Power Electronics and Design*, pp. 297-300, 1996.

[5.4] Q. Wang and S. Vrudhula, "Multi-level Logic Optimization for Low Power Using Local Logic Transformation," *Proceedings of International Conference of Computer-Aided Design*, pp. 270-277, 1996.

[5.5] B. Rohfleisch, A. Kolbl and B. Wurth, "Reducing Power Dissipation after Technology Mapping by Structural Transformations," *Proceedings of Design Automation Conference*, pp. 789-794, 1996.

[5.6] S. Chang, M. Marek-Sadowska and K. Cheng, "Perturb and Simplify: Multilevel Boolean Network Optimizer," *IEEE Transactions on Computer-Aided Design of Integrated Circuits and Systems*, vol. 15, no. 12, pp. 1494-1504, Dec. 1996.

[5.7] R. Panda and F. Najm, "Post-mapping Transformation for Low-Power Synthesis," To appear in *VLSI Design*.

[5.8] M. Stan and W. Burleson, "Bus-Invert Coding for Low Power I/O," *IEEE Transactions on VLSI Systems*, vol. 3, no. 1, pp. 49-58, Mar. 1995.

[5.9] E. Olson and S. Kang, "Low-Power State Assignment for Finite State Machines," *Proceedings of International Workshop on Low Power Design*, pp. 63-68, 1994.

[5.10] L. Benini and G. De Micheli, "State assignment for Low Power Dissipation," *Proceedings of Custom Integrated Circuit Conference*, pp. 136-139, 1994.

[5.11]   G. Hachtel, M. Hermida, A. Pardo, M. Poncino and F. Somenzi, "Re-Encoding Sequential Circuits to Reduce Power Dissipation," *Proceedings of International Conference of Computer-Aided Design*, pp. 70-73, 1994.

[5.12]   C. Tsui, M. Pedram, C. Chen and A. Despain, "Low Power State Assignment Targeting Two- and Multi-level Logic Implementations," *Proceedings of International Conference of Computer-Aided Design*, pp. 82-87, 1994.

[5.13]   V. Veeramachaneni, A. Tyagi and S. Rajgopal, "Re-encoding for Low Power State Assignment of FSMs," *Proceedings of International Symposium on Low Power Design*, pp. 173-178, 1995.

[5.14]   D. Chen, M. Sarrafzadeh and G. Yeap, "State Encoding of Finite State Machines for Low Power Design," *Proceedings of International Symposium on Circuits and Systems*, pp. 1309-1312, 1995.

[5.15]   G. Yeap, "CPU Controller Optimization for HDL Logic Synthesis," *Proceedings of Custom Integrated Circuit Conference*, pp. 127-130, 1997.

[5.16]   M. Alidina, J. Monteiro, S. Devadas, A. Ghosh and M. Papaefthymiou, "Precomputation-Based Sequential Logic Optimization for Low Power," *IEEE Transactions on VLSI Systems*, vol. 2, no. 4, pp. 426-436, Dec. 1994.

[5.17]   V. Tiwari, S. Malik and P. Ashar, "Guarded Evaluation: Pushing Power Management to Logic Synthesis Design," *Proceedings of International Symposium on Low Power Design*, pp. 221-226, 1995.

| CHAPTER 6 | *Special Techniques* |

In the previous chapters, we have discussed methods to reduce power consumption at various levels of design abstraction. The methods are general-purpose because they can be applied to wide varieties of circuits. We have focused on the general techniques of low power design with some specific examples to highlight their applications. In this chapter, we will present some special low power techniques that can be applied to certain types of circuits. The special techniques typically exploit certain unique properties of the circuit and often results in dramatic power reduction.

There is no clear distinction between "general" and "special" techniques. The difference is in the scope of their applications. The special techniques require more manual effort because design automation in these areas is less mature. Although the techniques are termed "special," their applications could be very important because the circuits are used in many VLSI chips; for example, the low power clock, bus and SRAM techniques. Despite the narrow scope of application, it is hoped that the techniques featured in this chapter could become a source of inspiration for the readers.

6.1 Power Reduction in Clock Networks

In a synchronous digital chip, the clock signal is generally one with the highest frequency. The clock signal typically drives a large load because it has to reach many sequential elements distributed throughout the chip. Therefore, clock signals have been a notorious source of power dissipation because of high frequency and load. It has been observed that clock distribution can take up to 40% of the total power dissipation of a high performance microprocessor [6.1]. Furthermore, the clock signal caries no information contents because it is predictable. It does not perform useful computation and only serves the purpose of synchronization. The number of different clock signals on a chip is very limited and warrant special attention during the design process. As such, many special techniques have been devoted to the power efficiency of clock generation and distribution.

6.1.1 Clock Gating

In Section 5.2, we discussed signal gating as a general means to mask unwanted signal transitions from propagating forward. The same idea has been applied to clock signals.

Clock gating, as depicted in Figure 6.1, is the most popular method for power reduction of clock signals. When the clock signal of a functional module (ALUs, memories, FPUs, etc.) is not required for some extended period, we use a gating function (typically NAND or NOR gate) to turn off the clock feeding the module. Note that the gating signal should be enabled and disabled at a much slower rate compared to the clock frequency. Otherwise the power required to drive the enable signal may outweigh the power saving. Clock gating saves power by reducing unnecessary clock activities inside the gated module.

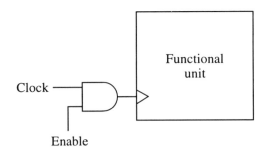

FIGURE 6.1 Clock gating.

Clock gating has been successfully used in recent high performance CPU designs [6.2]. One reason is that in a CPU chip, some functional modules may be left idle for an extensive period depending on the software it is executing, for example the floating point unit of a CPU. The design complexity and performance degradation of clock gating are generally manageable. The gated clock signal suffers an additional gate delay due to the gating function. In most high performance chip design, the clock distribution network consists of a multiple hierarchy tree feeding the sequential elements. The masking gate simply replaces one of the buffers in the clock distribution tree. Therefore, the delay penalty is generally not a concern. If the gating signal appears in a critical delay path and degrades the overall speed, the designer can always choose not to gate a particular module. The generation of enable/disable signals increases the complexity of control logic design. The timing relation of this signal has to be carefully evaluated to avoid signal glitches at the clock output. Often, it is harmless to allow the clock signal to pass through when the functional module is unused. During test mode, the gating signal may have to be removed to facilitate testing.

6.1.2 Reduced Swing Clock

Referring to the golden $P = CV^2 f$ equation, the most attractive parameter to attack is the voltage swing V due to the quadratic effect. Generally, it is difficult to reduce the load capacitance or frequency of clock signals due to the obvious performance reasons. We will discuss some power reduction techniques for clock signals based on voltage swing reduction.

One such idea proposes to reduce the clock swing by half [6.3]. Notice that in CMOS design, a clock signal is only connected to the gate of a transistor when it reaches a sequential element. The clock signal is seldom connected to the source or drain of a transistor. Inside a sequential cell, the clock signal is used to turn on or turn off transistors. Consider a 5V digital CMOS chip with an N-transistor threshold voltage of 0.8V. For a 5V regular full swing clock signal, an N-transistor gated by the clock will turn on if the clock signal is above 0.8V. Therefore if we limit the swing of the N-transistor clock signal from zero to 2.5V (half swing), the on-off characteristics of all N-transistors remain digitally identical. Similar observation can be made for the clock signal feeding a P-transistor, where the swing is limited from 2.5V to 5V. Figure 6.2 illustrates the clock waveform of the half-swing clocking scheme.

The half-swing clock is especially attractive in the two-phase non-overlapping clocking design where a sequential element requires two out-of-phase clock signals. The power saved from the reduced swing is 75% on the clock signal. The penalty incurred is the reduced speed of the sequential elements. The sequential delay, expressed in propagation delay and setup/hold time, is approximately doubled. This is because the on-resistance of a transistor is inversely proportional to the voltage difference between its gate and source and the reduced clock swing increases the on-resistance of the transistors. However, if the contribution of the sequential element delay to the

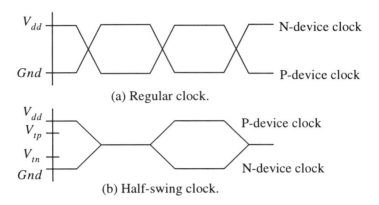

FIGURE 6.2 Conventional and half-swing clocking scheme.

total delay is relatively small, which is common in medium speed digital systems, the degradation to the overall system performance may be tolerable.

Generating the half-swing clock signal is also relatively simple using the charge sharing principle with stacked inverters. Figure 6.3 is a circuit to generate the clock waveforms. The capacitance C_1, C_2, C_3 and C_4 are the parasitic capacitances of the circuit. C_A and C_B are add-on capacitance much larger than the parasitic capacitance. From the principle of charge sharing, when CLK is low, the voltage at V_H is

$$V_H = \frac{C_1 + C_A}{C_1 + C_4 + C_A + C_B} V_{dd} \tag{6.1}$$

Similarly, when CLK is high, the voltage is

$$V_H = \frac{C_2 + C_A}{C_2 + C_3 + C_A + C_B} V_{dd} \tag{6.2}$$

When we set $C_A = C_B \gg C_1, C_2, C_3, C_4$, the voltage V_H will be clamped at $0.5V_{dd}$. The circuit generates two-phase, half-swing waveforms. The signals $CKPT$ and $CKPB$ are used to clock the P-transistors while $CKNT$ and $CKNB$ are for the N-transistors. Creating large on-chip capacitance C_A and C_B is difficult and may require an external pin for the node V_H. A simpler charge sharing circuit is shown in Figure 6.4. This circuit relies on the parasitic loading of the clock lines C_p and C_n to achieve the charge sharing effect. The capacitance C_p and C_n should be equalized to obtain the proper half-swing waveform. The circuit is simpler but more susceptible to process variation than the one in Figure 6.3.

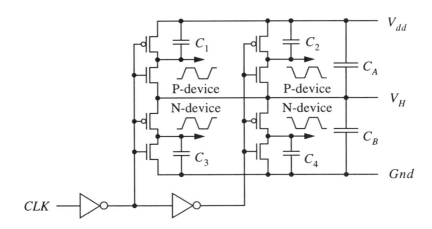

FIGURE 6.3 Charge sharing circuit for half-swing clock.

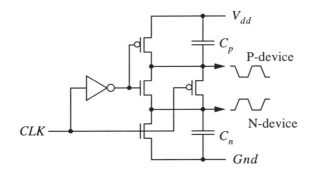

FIGURE 6.4 Simple charge sharing circuit for half-swing clock.

Perhaps the most obvious advantage of the half-swing clock scheme is its simplicity. Only the clock distribution circuit needs to be specially crafted to support the half-swing clock. The sequential and combinational elements of the circuit remain unchanged throughout the design.

As with any circuit that relies on charge sharing, a minimum clock frequency needs to be maintained to avoid charge leakage. This makes it impossible to hold the clock generator for an extended period to implement sleep modes. Also, for low voltage chips, e.g., $V_{dd} = 1.0V$, $V_{tn} = 0.4V$, the reduced clock swing probably cannot sustain acceptable delay on the sequential elements because the gate voltage is only

0.1 V above the threshold. This limits the usefulness of the half-swing clock scheme in very low voltage systems.

6.1.3 Oscillator Circuit for Clock Generation

Some chips contain on-chip clock generation circuits using crystal oscillator devices. The simplest oscillator circuit is shown in Figure 6.5. The inverter I_0 serves as a gain amplifier for the feedback oscillator and the inverter I_1 reshapes the waveform to obtain a proper digital clock signal. To minimize power dissipation of the circuit, several trade-off decisions have to be considered. For good stability of oscillation, we like to set a large gain on the inverter I_0. However, a large gain results in large voltage swing and short-circuit current that increase the power dissipation, as shown in Figure 6.5. Thus, it is important to tune the transistor sizes of the inverters to achieve a proper trade-off between the "quality" of oscillation and the power dissipation of the circuit.

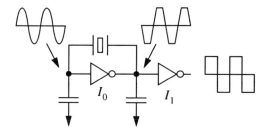

FIGURE 6.5 An on-chip crystal oscillator circuit.

To reduce the short-circuit current, resistors can be placed in series with the transistors of the inverter I_0. We can also use a dual voltage system where the oscillator circuit is operated at a lower voltage. This requires level shifters before the clock signal is delivered to its destinations. For more advanced designs, an automatic-gain-controlled amplifier is used to provide better voltage swing control, oscillation stability and process variation tolerance. Although a low gain amplifier is good for power dissipation, it does lengthen the start-up time of the oscillator circuit. One way to improve start-up time is to increase the gain during power-up and reduce it after reaching a stable oscillation.

6.1.4 Frequency Division and Multiplication

A power reduction scheme that has been successfully applied is frequency division and multiplication shown in Figure 6.6. This is especially common for off-chip clock signals because they drive very large capacitance. The off-chip clock signal runs at a slower speed and an on-chip phase-locked loop circuit is used to multiply the frequency to the desired rate. The slower signal also eases the off-chip signal distribution in terms of electromagnetic interference and reliability. The frequency multiplier N is a trade-off between power dissipation and the phase-locked loop circuit complexity. Larger values of N lead to better power dissipation but increases the design complexity and performance of the phase-locked loop circuit.

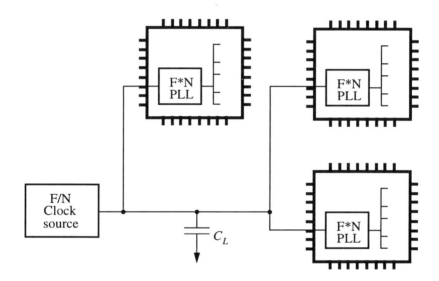

FIGURE 6.6 Frequency division and multiplication.

6.1.5 Other Clock Power Reduction Techniques

A well-known scheme used in older microprocessors is called clock masking or clock stretching. It involves inhibiting clock pulses to the processor when it is not needed, for example when the processor is waiting for memory or I/O devices. This has been applied to early generations of portable computers. This technique is probably obsolete today because modern CPUs now have high speed cache memories that eliminate idle cycles.

Yet other techniques attempt to reduce the capacitance of the clock signal. In some advanced microprocessor chips, the clock signal is routed on the topmost metal layer, thus reducing the capacitance [6.1]. In [6.4], the gated modules are clustered based on their activity patterns to reduce power dissipation and complexity of control signal generation. Simultaneous switching of global clock signals can also cause large transient current to be drawn. Methods to reduce the transient current by controlling clock skew have been studied [6.5].

6.2 CMOS Floating Node

In a low power CMOS chip design, special precautions must be taken to make sure that all nodes are at the full logic level, i.e., either V_{dd} or *Gnd*. This means that all nodes should be actively driven by some gates. A node is called *floating* if it is not driven by any gates, i.e., all transistors connected to the node are turned off. When a node is left floating for a long time, charge decay and noise coupling can cause the node voltage to reach intermediate values. This is extremely harmful to a low power system. Any CMOS fanin gate connected to the node will have a large short-circuit current when its input is between V_{tn} and $V_{dd} - V_{tp}$ because both P and N transistors are turned on. In severe cases, this may cause the gate to overheat and fail permanently. Sometimes, the noise at the floating node may be amplified by the downstream logic gates and cause unnecessary switching activities. Floating nodes are easy to avoid with careful analysis during the logic and circuit design.

6.2.1 Tristate Keeper Circuit

The floating node situation occurs frequently in bus design. Busses are mostly driven by multiple tristate drivers located at various parts of a chip. If all tristate drivers of a bus are turned off for an extended period, the bus is essentially left floating. This can occur during sleep mode or partial power down mode. Many designers simply put a weak pull-up (or pull-down) transistor (called a bleeder transistor) that is constantly turned on, as shown in Figure 6.7(a). This solves the floating node problem but introduces a short-circuit path from V_{dd} to *Gnd* through the keeper transistor when the bus is at logic 0. Better floating suppressor circuits are shown in Figure 6.7(b) and (c) using cross coupled inverters. These circuits ensure that there is no steady short-circuit current path through the keeper transistors regardless of the logic state. The output transistors of the keeper circuits must be weak enough so that the signal drivers can overpower the keeper for proper logic operations. Typically, long channel transistors are used.

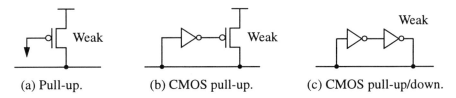

(a) Pull-up.   (b) CMOS pull-up.   (c) CMOS pull-up/down.

FIGURE 6.7 Floating bus suppressor circuit.

The keeper circuits slightly increase the power dissipation and delay of the bus during normal operations. If a bus is not left floating for a long time (e.g., less than 1ms), the keeper circuit may not be required. However if a chip can enter sleep modes, during which the busses are not driven, the keeper circuit is essential. Most often, the additional power dissipation of the keeper circuits is negligible.

6.2.2 Blocking Gate

The floating node problem can also occur at hibernation mode in which some parts of a chip or system are powered-down. In this case, some nodes become floating because its driver gate is not powered. However, the fanin gates connected to the floating nodes can experience large short-circuit current if not properly designed. One easy way to solve this problem is to use a blocking NAND gate as shown in Figure 6.8(a). At hibernation mode, the enable signal is set to logic 0 so that the floating node does not cause short-circuit current.

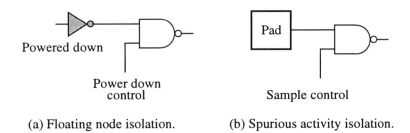

(a) Floating node isolation.   (b) Spurious activity isolation.

FIGURE 6.8 Blocking NAND gate.

The blocking gate technique is typically used when the number of fanin gates of the floating node is small because we need one blocking gate for each fanin gate. For a bus with many fanin gates, a keeper circuit is more appropriate since only one keeper

147

circuit is needed. Unlike the bus keeper circuit, the blocking gate requires a control signal for proper operation.

Besides blocking a floating node, the gating technique is also useful for blocking unwanted signal transitions, thus saving power. For example, at the data-input pin of a chip, the blocking gate can be used to prevent spurious transition from entering the chip before the data is ready to be sampled. Figure 6.8(b) shows such application of blocking gate. This is the same as the signal gating concept already mentioned in Section 5.2.

6.3 Low Power Bus

In digital systems, busses serve as communication medium that delivers data to various parts of a chip. Busses are significant sources of power dissipation in a digital system due to high switching activities and large and capacitance. Advanced systems are typically characterized by wide and high speed busses which means that the capacitance and frequency factor of the power dissipation dominates.

In Section 5.3.2, we discussed the bus invert encoding method to reduce the average and maximum toggle activities of a bus. In this section, we will see more techniques for reducing power dissipation of a bus. All techniques discussed here rely on reduced voltage swing to achieve the low power goal.

6.3.1 Low Swing Bus

Because a bus contains multiple bits, differential signaling using low voltage swing can be implemented with relatively low overhead. This is generally not desirable for an individual signal because the signal wire has to be doubled. Busses also have the unique property that all signals in the bus operate in sync. This leads to some very creative circuit techniques for differential signaling.

The simplest differential signaling scheme for buses is to introduce an additional reference voltage line between the signal sender and the receiver units. With the reference voltage, the voltage swing of each signal bit can be reduced. An early version of this idea was proposed in [6.6]. The circuit technique was later enhanced in [6.7]. The basic architecture and signaling scheme of the bus are shown in Figure 6.9. Figure 6.10 shows a circuit implementation of the bus driver and receiver. Notice that the bus receiver circuit is very similar to the high-speed latch shown in Figure 4.12.

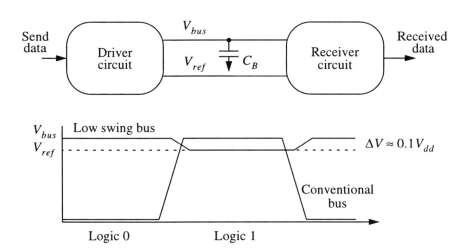

FIGURE 6.9 Low swing bus system and signals.

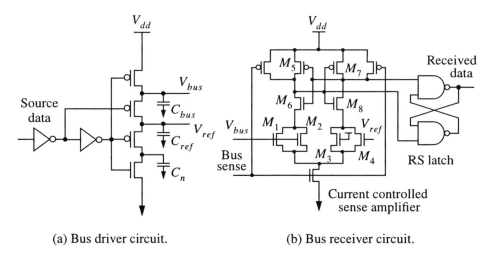

(a) Bus driver circuit.　　　　　　(b) Bus receiver circuit.

FIGURE 6.10 Low swing bus circuit techniques.

The bus driver circuit relies on charge sharing to produce the reduced bus swing and the reference voltage. Consider the driver circuit in Figure 6.10(a). We like to generate a constant reference voltage V_{ref} at C_{ref}. This requires that $C_{ref} \gg C_n, C_{bus}$ so that V_{ref} is relatively stable when charge sharing occurs. When the input is at logic 1, the charge at C_n is grounded and the charge lost is $C_n V_{ref}$. At the same time, C_{bus} is charged to V_{dd}. When the input is at logic 0, C_{bus} is at V_{ref} and the charge transferred to C_{ref} is $C_{bus}(V_{dd} - V_{ref})$. At equilibrium the charge entering and leaving C_{ref} should equal and we have

$$C_n V_{ref} = C_{bus}(V_{dd} - V_{ref}) \qquad (6.3)$$

$$V_{ref} = \frac{C_{bus}}{(C_n + C_{bus})} V_{dd} \qquad (6.4)$$

The capacitance C_n should be tuned to provide the proper V_{ref} value to achieve the desired swing reduction.

The receiver circuit uses differential sensing to detect the data signal on the bus. A *bus sense* signal is used to disable and precharge the receiver circuit when the bus signal is not stabilized. During sensing, the differential amplifier evaluates the bus signal and the cross coupled NAND gates latch the data until the next precharge cycle. The differential transistors M_1, M_2, M_3 and M_4 have to be matched. When the bus value is at logic 0, V_{bus} is at V_{dd} and the left transistor pair conducts more than the right pair. At logic 1, V_{bus} is at V_{ref} and the right transistor pair conducts more. The current difference produced by the differential pairs of transistors topples the cross coupled inverter pair (M_5, M_6), (M_7, M_8) which is at a metastable state after precharge. Unlike traditional differential sensing circuits that typically consume DC current and defeat the low power purpose, the CMOS style cross coupled inverter used in this circuit eliminates the DC current after the completion of sensing.

The power efficiency of the bussing scheme depends on the extent of voltage swing reduction, which in turn, depends on the noise immunity and sensitivity of the differential sensing circuitry. In [6.7], a reduced swing of $0.1 V_{dd}$ is suggested, which leads to an impressive 99% switching power saving compared to a full swing bus. Of cause the additional power dissipated at the driver and the receiver circuits need to be considered. The power efficiency becomes more favorable when the bus capacitance is larger. Additional timing delays are incurred in the driver and the receiver circuits but the reduced swing helps to improve the signal switching time. Also the reduced voltage swing leads to smaller transistors at the driver circuits and helps to reduce the short-circuit current. The application of this technique to off-chip busses is very attractive because of large bus capacitance and relatively small overhead in the additional circuitry.

There are some limitations with the bussing scheme. The reduced voltage swing is susceptible to noise interference and cross-talk between adjacent bit lines. Also, the design of the differential sensing circuit may pose difficulties at very low operating voltage for aggressive low power systems. The voltage swing reduction depends on the ratio of C_{bus} and C_n, which may be difficult to control to a desired limit. Since producing large on-chip capacitance is difficult, the capacitance C_{ref} may have to be installed off-chip.

Because the bus driver circuit uses charge sharing, it is inherently a dynamic circuit. It cannot stay dormant for a long period due to charge leakage. This may be a problem if the data signal contains long series of identical data value. If the bus circuit is deactivated, it takes some time for the reference voltage to recover to the proper operating level.

Other low swing bus signaling techniques have been reported. In [6.6], the bus signal is discharged to a *dummy ground* that serves as a charge storage capacitor and produces variable differential swing depending on the number of data bits being discharged. The variable differential swing requires some changes in the differential sensing circuit. However, the power efficiency of the scheme is relative insensitive to process variation.

6.3.2 Charge Recycling Bus

The low swing bus presented in the previous section discharges to a common reference signal line. In this section, we present another bussing technique in which a high order data bit discharges to the lower bit. This essentially "recycles" the charge for reuse at the lower bit. However, two wires per bit are required to achieve this effect.

The circuit principle and the bus signals are illustrated in Figure 6.11 [6.8]. Each data bit i has two wires $Di+$ (solid lines) and $Di-$ (dashed lines) which provide differential signals to the bus receiver. At the precharge phase, the switches S_1 are closed and S_2 are opened so that the voltage at capacitors C_{Di+} and C_{Di-} are equalized. Depending on the logic value of the bus bit, a parallel or crossed connection is realized at the logic evaluation module. At the evaluation cycle, switches S_2 are closed and S_1 are opened. One line from a high order bit is equalized with another line of a low order bit through the switch S_2. This produces differential voltage at each line pair, as shown in Figure 6.11(b), depending on the logic value of the bit. At the end of the evaluation phase, the signal pair is again equalized in the next precharge phase.

An actual circuit that implements the switches and the logic evaluation module is shown in Figure 6.12. Depending on the logic value of a bit, the transistor pair (M_1, M_4) or (M_2, M_3) is turned on during the evaluation cycle. At precharge, M_1, M_2, M_3 and M_4 are turned off and M_5 is turned on to allow charge equalization.

For proper operation, PMOS transistors have to be used at the top-half of the bus driver circuit and NMOS transistors at the bottom-half. This is necessary to prevent a

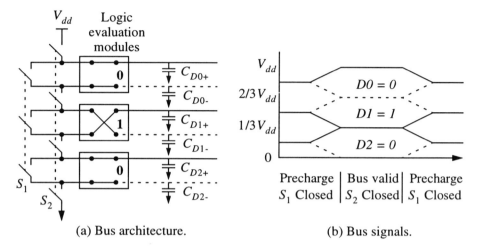

(a) Bus architecture.　　　　　(b) Bus signals.

FIGURE 6.11 Charge recycling bus.

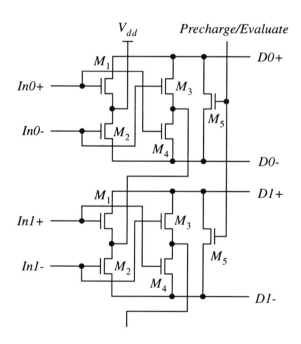

FIGURE 6.12 Driver circuit implementation of charge recycling bus.

threshold voltage drop when the transistors are turned on. The bus receiver circuit uses differential voltage sensing technique like the one described in the previous section. The proposed receiver circuit is shown in Figure 6.13. Again, two types of receiver circuit, using PMOS or NMOS transistors have to be used depending on the bit position of the receiver.

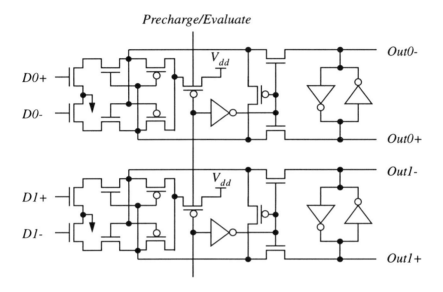

FIGURE 6.13 Receiver circuit of the charge recycling bus.

The voltage swing reduction of the charge recycling bus depends on the number of bits stacked. For an n-bit bus, the voltage swing of each bus line is

$$\Delta V = \frac{1}{2n} V_{dd} \tag{6.5}$$

which leads to a differential signal of $(1/n) V_{dd}$ for each bus pair. We analyze the efficiency of the bussing scheme by using a conventional n-wire full swing bus as the baseline. The conventional bus has n capacitors, each with capacitance C and a voltage swing of V_{dd} operating at frequency f. The power consumption of the conventional bus is

$$P_{conv} = nCV_{dd}^2 f \tag{6.6}$$

The charge recycling bus has $2n$ capacitors each with a voltage swing of $V_{dd}/(2n)$. Since the bus requires precharge and evaluate phases, the capacitors are charged and discharged twice in a complete bus cycle, which essentially double the frequency of charging and discharging for the same bus throughput. The power dissipation of the charge recycling bus P_{crb} is therefore given by

$$P_{crb} = 2nC\left(\frac{V_{dd}}{2n}\right)^2 (2f) = \frac{1}{n^2}P_{conv} \qquad (6.7)$$

Notice that in a conventional bus, the power dissipation is data dependent because charging or discharging is not required if the current bit value agrees with the previous bit value. However, in the charge recycling bus, all switching occurs despite the time correlation of the data bits. If we assume a uniform random probability on each data bit, the conventional bus only switches half the bits on the average. Thus, the efficiency of the charge recycling bus is reduced to

$$P_{crb} = \frac{2}{n^2}P_{conv} \qquad (6.8)$$

The quadratic increase in efficiency with the bus width n is the result of squaring effect of the voltage swing reduction. The acceptable bus stacking depends on the noise immunity and the sensitivity of the receiver circuit. A large value of n implies a smaller voltage swing. This improves power efficiency but decreases noise immunity.

In this bus signaling scheme, maintaining all line capacitances of the bus to an equal value is important. Otherwise, the waveform may fluctuate from the ideal value causing noise margin loss. This somewhat limits the application of the scheme to on-chip buses only because equalizing off-chip capacitances is difficult. Special precautions need to be taken with the layout of the bus to balance the capacitances. For example, if the bus makes a turn, the outer bit lines will have larger capacitance than the inner lines because they are longer. This may be compensated by another turn in the opposite direction or other balancing technique.

An experimental prototype of the bussing scheme has been demonstrated in a laboratory setting [6.8]. The bus reportedly achieved clock frequency of 50MHz at the operating voltage of 3.0V with 4-bit stacking (0.75V differential) using a standard 0.5um CMOS process.

The charge recycling bus scheme requires a precharge phase for proper operation. During precharge, no data are transmitted on the bus, reducing the data transfer rate of the bus. The same inventors also proposed a time-multiplexed version of the bussing scheme [6.9]. In this scheme, two sets of identical bus drivers are allocated to a differential wire pair. When one bus driver is sending data to the signal pair, the other driver is in precharge phase and vice versa. This essentially doubles the data transfer capacity with a modest increase in the circuit complexity. The time-multiplexed bussing scheme is shown in Figure 6.14.

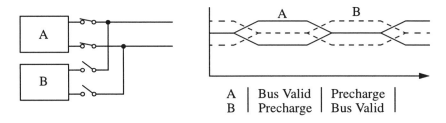

FIGURE 6.14 Time multiplexed charge recycling bus.

6.4 Delay Balancing

A major problem in designing a low power parallel multiplier is the glitch power. A parallel multiplier is a series of cascaded adders with long carry chains. Because of the unbalanced nature of the parallel multiplier architecture, the output signals of the adder are very susceptible to glitches.

To illustrate the glitch problem, we consider the architecture of a simple array multiplier shown in Figure 6.15. For simplicity, we assume a *unit delay model* where the delay through each multiplier cell from the sum or carry input to the sum or carry output is one. The input signals of the multiplier arrive simultaneously. After one delay unit, the outputs of all multiplier cells M_{ij} may potentially switch in response to the input transitions. After two delay units, all cell outputs may switch except the outputs of M_{00}. At three delay units, the outputs of M_{01} stop switching; etc. In short, the outputs of a multiplier cell may switch before $n + 1$ delay units if the longest path from its outputs to the primary inputs is n. The longest path lengths of the outputs of cells $[M_{03}, M_{02}, M_{01}, M_{00}]$ are $[4, 3, 2, 1]$; cells $[M_{13}, M_{12}, M_{11}, M_{10}]$ are $[6, 5, 4, 3]$; etc. The longest path of a multiplier cell starts from the sum or carry output; follows the carry chain *rowise* until the rightmost cell of the row; then follow the sum input to the upper row. At each row, it always follows the carry chain before it has to move to a higher row at the end of the chain. Thus, cell M_{ij} has the longest delay path length of $2i + j + 1$. It can be seen that the outputs of most multiplier cells have many signal glitches because they can make many switches before settling to the final value.

155

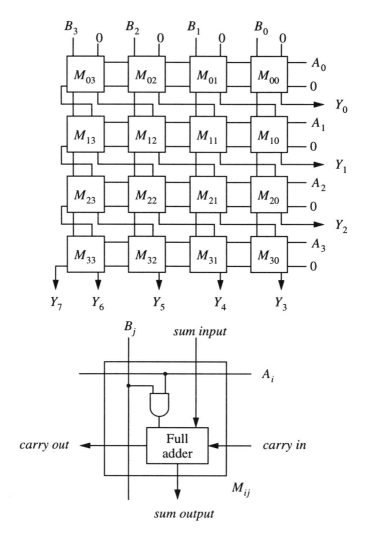

FIGURE 6.15 Architecture of an array multiplier.

Consider the structure of the multiplier in detail. The right adder-input of cell M_{ij} obtains its value from cell $M_{(i-1)(j+1)}$. The carry input of M_{ij} obtain its value from $M_{i(j-1)}$. The longest delay of $M_{(i-1)(j+1)}$ and $M_{i(j-1)}$, which happens to be identical, is $2i+j$ delay units. Thus, the carry and sum input signals of M_{ij} arrive at $2i+j$ delay units. The left adder-input of a multiplier cell obtains its value immediately from the primary input of the multiplier (A_i and B_j), which has a zero delay (ignore the AND-gate delay). If we can *intentionally* delay the arrival time of the left adder-input of M_{ij} by $2i + j$ units, we will have all inputs of M_{ij} arriving simultaneously and its output glitches will be reduced. There is an exception at column 3 because the input is obtained from the cell immediately above it. In this case, the right adder-input needs to be delayed by one unit.

The delay elements for the left adder-input can be constructed in a distributed manner. Because the carry and sum input signal arrival time of M_{ij} is $2i + j$ units, if we move from the right to the left within a row, the delay requirement increases by one unit. If we move from the top to the bottom within a column, the delay requirement increases by two units. We can use these properties to embed the delay lines in the multiplier cells. The configuration of the delay lines in the multiplier cells is shown in Figure 6.16. The multiplier cells at the topmost row and rightmost column will require different delay values.

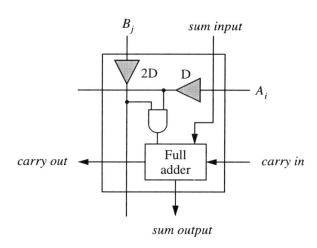

FIGURE 6.16 Delay balanced multiplier cell.

The delay-balancing technique can be used on all types of parallel array multipliers such as Booth's multiplier and Wallace Tree multiplier. The difference is that in these multiplier architectures, the delay analysis is more involved. The technique has been reported in [6.10] and achieves power reduction of 36% on a parallel array multiplier.

6.5 Low Power Techniques for SRAM

Memory devices, whether dynamic, static or read-only, are found in many digital chips. Increasing packing density has allowed larger RAMs to be integrated into a single chip. Many advanced processors now have on-chip instruction and data caches using high speed SRAMs. The RAM contents of a chip have increased dramatically over the years and the trend continues. Thus, improving the power efficiency of an SRAM is critical to the overall system power dissipation.

In any low power application, reducing operating voltage is always a good method to lower dynamic power dissipation. However, the performance and other system requirements typically limit the operating voltage and we have to look for other techniques. Since SRAMs have relatively fixed architecture, the dynamic power reduction techniques are easily classified. An SRAM consists of RAM cell arrays, row decoders, sense amplifiers, column decoders and some peripheral circuits as shown in Figure 6.17.

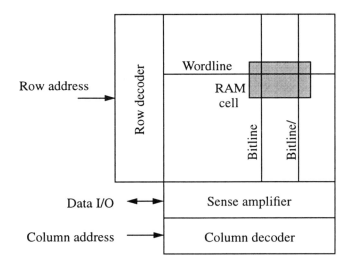

FIGURE 6.17 Architecture of an SRAM.

The low power SRAM design has two facets: standby power reduction and operating power reduction. The relative importance of the power reduction techniques depends on the application. In most battery power system, the SRAM typically spends most of its time in hibernation while in high performance systems the standby power may not be as important.

6.5.1 SRAM Cell

In some SRAM design, a resistor load (typically using a poly layer) is used, forming a 4-transistor cell. This makes efficient use of the area but there is a constant static current that flows through the cell. As SRAM size increases, the static current needs to be decreased; otherwise, the total static current of the SRAM becomes intolerable. The static current has to be much larger than the leakage current of the transistor for the proper operation of the 4-transistor cell. Eventually the 4-transistor cell design reaches a limit because the static current cannot be further decreased.

An SRAM cell using 6-transistors (PMOS loads) can reduce the static current to the order of the leakage current of CMOS transistors but it takes more area for the cell layout. The reduction in static current from a resistor load to a PMOS load could be several hundred times. Thus for low power SRAM design, the 6-transistor cell is typically used. The 4-transistor and 6-transistor circuits are shown in Figure 6.18.

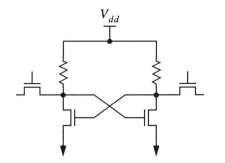
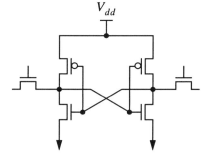

(a) 4-T cell with resistive load.   (b) 6-T cell with PMOS load.

FIGURE 6.18 SRAM cell circuits.

In aggressive low power design, high density RAMs are required to operate at very low voltage level to reduce power dissipation. The threshold voltage of the device is often reduced to compensate for the speed degradation due to low operating voltage. The reduction of threshold voltage leads to exponential increases in static leakage current of the SRAM cell (see Section 1.4) and could become a limiting factor. One solution is to use multiple threshold devices in the memory chip. For example, the memory cells use higher threshold voltage devices to reduce leakage while the rest of the peripheral circuits use low threshold devices to improve speed (see Section 4.6). A more exotic solution is to adjust the threshold voltage dynamically by adjusting the back bias voltage of the transistor. When a memory bank is activated, its threshold voltage is reduced for speed and during standby, the threshold voltage is increased to reduce leakage. This technique has also been successfully used in some DRAM chips.

6.5.2 Memory Bank Partitioning

A well-known technique used to improve speed is to partition a memory array into smaller memory banks so that only the addressed bank is activated. Interestingly, this technique can also improve the power efficiency of the SRAM because the wordline capacitance being switched and the number of bit cells activated are reduced. The partitioning can be applied hierarchically to produce smaller memory banks. However, at some point the delay and power overhead associated with the memory bank decoding circuit dominates and further partitioning becomes harmful. Typically, two to eight bank partitions are appropriate.

6.5.3 Pulsed Wordline and Reduced Bitline Swing

In a conventional SRAM design, the bitlines are allowed to swing from rail to rail during a read operation. To conserve power, the voltage swing at the bitlines needs to be limited. One way to achieve this is to isolate the memory cells from the bitlines after a successful differential sensing. This prevents the memory cells from changing the bitline voltage further. In some cases, the bitline sense amplifiers also need to be isolated from the bitlines after sensing. This prevents the bitline capacitance from having large voltage swing, thus saving power. This method requires pulsed signals at the wordlines and the sense amplifiers.

The generation of wordline pulses is very critical because if the pulses are too short, the sense operations may fail. On the other hand if the pulses are too long, the power efficiency is degraded because the bitline swing depends on the duration of the pulses. A simple technique is to use delay lines to generate pulses with a fixed width. However, delay lines are susceptible to process, temperature and voltage variation. A better method is to use a feedback mechanism from the bitlines. Since the goal of pulsing the wordlines is to limit the bitline swing, a bitline swing detector can be constructed on an additional dummy cell column [6.11]. The height of the dummy column is set to 10% of the regular column and the cells are tied to a fixed value. This means that the capacitance of the dummy column is only 10% of the capacitance of a regular column. Thus, when the regular bitlines reach 10% swing, the dummy bitlines have reached full swing and triggers the pulse shut-off. This limits the bitline voltage swing to a precise range with much less sensitivity to process variation, temperature and operating voltage. The circuit technique is shown in Figure 6.19.

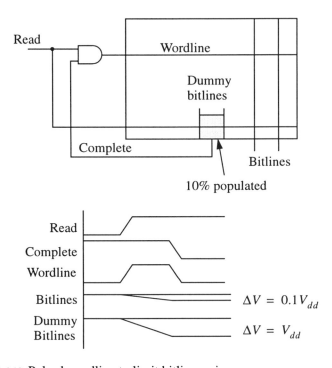

FIGURE 6.19 Pulsed wordline to limit bitline swing.

The swing limiting technique can also be applied to any part of the bitline datapath with large capacitance. Most pulsed wordline techniques increase the delay and design complexity of the SRAM. If the write-power of the SRAM is a concern, the voltage swing of the bitlines during the write cycle can also be limited, as we will see in the next section.

In some SRAM sense amplifier design, the amplifier may swing the bitlines after the cell access transistors have been turned off. This defeats the intent of the pulsed word-line implementation, which is to reduce bitline swing. A method to solve this problem is to use bitline isolation technique [6.12] as shown in Figure 6.20. The *sense* signal is asserted to initiate the read sense amplifier. After a sufficient differential voltage has developed on the bitlines, the *isolate* signal is asserted to isolate the sense amplifier from the bitlines. This prevents the sense amplifier from further swinging the bitline capacitance. The isolate and sense signals need to be timed properly to balance the trade-off among noise immunity, power dissipation and sensing delay.

161

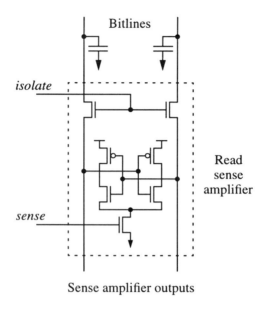

Bitlines

isolate

Read
sense
amplifier

sense

Sense amplifier outputs

FIGURE 6.20 Bitline isolation to reduce bitline swing.

6.5.4 Case Study: Design of an FIFO Buffer

In this section we present an exemplary low power design technique for an SRAM [6.13]. The intent is not to recommend the particular low power design technique but to highlight the chain of design decisions that exploits the special situations of the design domain. This leads to the ultimate low power goal in the particular operating environment of the SRAM. In fact, no single design technique can be applied in all scenarios. It is the job of a designer to weigh his design goals and constraints carefully, consider the available options and applies the necessary trade-off to reach the best design choices.

The particular SRAM in our case study was to be used as an FIFO (first-in-first-out) buffer in a network switching application. It has very wide words (more than 400 bits) and requires low power consumption. Generally, an SRAM is expected to have a much higher read-to-write ratio, especially those used as CPU caches. Therefore, the low power design techniques are concentrated in the read cycles. However, the access pattern of the FIFO buffer differs from that of an ordinary SRAM because the read-to-write ratio is 1:1. This means that besides the read cycle, the write cycle energy needs to be minimized as well.

162

The design is a standard 5V system with a 6-transistor CMOS SRAM cell. The designers used the pulsed wordline technique (see Section 6.5.3) to reduce the bitline swing during read and write cycles. Since there were 50% write cycles, the designer decided to reduce the bitline differential voltage during the *write* cycles to (0V, 1V). During the precharge cycles for read, the bitlines are precharged to 0.5V. In the sensing stage of the read cycles, the bitline voltage swing is limited by disabling the access transistors as soon as the bitline voltage difference is sufficient for sensing. The bitline voltage swing is thus limited to less than 1V during operation. Since bitline capacitance is large and the SRAM has a wide word length, the power reduction from the limited read and write voltage swing technique is substantial.

The use of the non-traditional (0V, 1V) write-voltage on the bitlines posed a problem in the SRAM cell design. Consider a 6-transistor CMOS RAM cell shown in Figure 6.21(a). In their reference design with (0V, 5V) differential write-voltage, the transistor sizes are given in Figure 6.21(b). In an SRAM cell design, the transistor size ratio $\beta_w/(\beta_n + \beta_p)$ needs to be kept below a certain threshold to maintain the read stability. Otherwise, noise at the bitline may corrupt the SRAM cell during the read cycles. The reduction of write differential voltage to (0V, 1V) prompted the designer to reduce β_n in order to guarantee successful write operations. However the reduced β_n caused read instability because the $\beta_w/(\beta_n + \beta_p)$ ratio was substantially increased. The ingenious designer solved the problem by reducing the wordline voltage of the access transistors from 5V to 3V during *read*. This essentially restored the read stability lost by the skewed $\beta_w/(\beta_n + \beta_p)$ ratio. The wordline voltage remained at 5V during the write cycles. The 3V wordline voltage and 1V bitline voltage were generated using charge pump circuits that consume additional power and increased the circuit design complexity. However, the overall benefit of reducing the write power was justifiable.

Let's revisit the chain of decisions in the design of the low power FIFO buffer. First the operation characteristics dictated that 50% of the SRAM accesses were write cycles. This led to a reduced bitline differential voltage from (0V, 5V) to (0V, 1V) during the write cycles to reduce the write power. The suppressed bitline voltage resulted in reduced transistor size β_n in the SRAM cell to allow successful write operations. This caused a read instability problem in the SRAM cell. The problem was eventually solved by reducing the wordline voltage to 3V during the read operations and a regular 5V wordline voltage during the write operations. The usual pulsed wordline technique was used to limit bitline swing during read cycles. This kind of decision chain is commonly encountered in low power design. A successful low power designer should familiarize himself or herself with such decision derivation and deduction process.

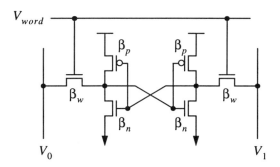

READD: $V_{word} = 5V$, $V_0 = 0V$, $V_1 = 1V$

WRITE: $V_{word} = 3V$, $V_0 = V_1 = 0.5V$

(a) CMOS SRAM cell and operation voltages.

Parameter	Ref. design (0V, 5V)	LP design (0V, 1V)
β_p	$50uA/V^2$	$50uA/V^2$
β_n	$365uA/V^2$	$108uA/V^2$
β_w	$108uA/V^2$	$108uA/V^2$
Write-energy	$25pJ$	$4pJ$

(b) Write-energy of conventional design versus reduced write differential design.

FIGURE 6.21 Low write energy memory cell design.

REFERENCES

[6.1] D. Dobberpuhl, *et al.*, "A 200-MHz 64-b Dual-Issue CMOS Microprocessor," *IEEE Journal of Solid-State Circuits*, vol. 27, no. 11, pp. 1555-1566, Nov. 1992.

[6.2]   G. Gerosa, *et al.*, "A 2.2W 80MHz Superscalar RISC Microprocessors," *IEEE Journal of Solid-State Circuits*, vol. 29, no. 12, pp. 1440-1454, Dec. 1994.

[6.3]   H. Kojima, S. Tanaka and K. Sasaki, "Half-Swing Clocking Scheme for 75% Power Saving in Clocking Circuitry," *IEEE Journal of Solid-State Circuits*, vol. 30, no. 4, pp. 432-435, Apr. 1995.

[6.4]   G. Tellez, A. Farrahi and M. Sarrafzadeh, "Activity-driven Clock Design for Low Power Circuits," *Proceedings of International Conference of Computer-Aided Design*, pp. 62-65, 1995.

[6.5]   P. Vuillod, L. Benini, A. Bogliolo and G. De Micheli, "Clock-skew Optimization for Peak Current Reduction," *Digest of Technical Papers, International Symposium on Low Power Electronics and Design*, pp. 265-270, 1996.

[6.6]   M. Hikari, H. Kojima, *et al.*, "Data-Dependent Logic Swing Internal Bus Architecture for Ultralow-Power LSI's," *IEEE Journal of Solid State Circuits*, vol. 30, no. 4, pp. 397-402, Apr. 1995.

[6.7]   A. Bellaouar, I. Abu-Khater and M. Elmasry, "An Ultra-Low-Power CMOS On-chip Interconnect Architecture," *Digest of Technical Papers, IEEE Symposium on Low Power Electronics*, pp. 52-53, 1995.

[6.8]   H. Yamauchi, H. Akamatsu and T. Fujita, "An Asymtotically Zero Power Charge-Recycling Bus Architecture for Battery-Operated Ultra-high Data Rate ULSI's," *IEEE Journal of Solid-State Circuits*, vol. 30, no. 4, pp. 423-431, Apr. 1995.

[6.9]   H. Yamauchi and A. Matsuzawa, "A Signal Swing Suppressing Strategy for Power and Layout Area Savings Using Time-Multiplexed Differential Data Transfer Scheme," *IEEE Journal of Solid-State Circuits*, vol. 31, no. 9, pp. 1285-1294, Apr. 1995.

[6.10] T. Sakuta, W. Lee and P. Balsara, "Delay Balanced Multipliers for Low Power/Low Voltage DSP Core," *Digest of Technical Papers, IEEE Symposium on Low Power Electronics*, pp. 36-37, 1995.

[6.11] B. Amrutur and M. Horowitz, "Techniques to Reduce Power in Fast Wide Memories," *Digest of Technical Papers, IEEE Symposium on Low Power Electronics*, pp. 92-93, 1994.

[6.12]   Y. Shimazaki, "An Automatic Power-Save Cache Memory for Low Power RISC Processors," *Digest of Technical Papers, IEEE Symposium on Low Power Electronics*, pp. 58-59, 1995.

[6.13]   J. Alowersson and P. Andersson, "SRAM Cell for Low-Power Write-in Buffer Memories," *Digest of Technical Papers, IEEE Symposium on Low Power Electronics*, pp. 60-61, 1995.

CHAPTER 7

Architecture and System

Today, the most common abstraction level for manual design capture is at the so called *register transfer level*. At this level, all synchronous registers, latches and the combinational logic between the sequential elements are described in a hardware description language such as *Verilog* or *VHDL*. The description is then transformed into logic gate implementations based on a technology specific cell library, a procedure known as logic synthesis. This allows a designer to handle more complex design because detailed manual analysis need not be made at the logic gate level that is more tedious and error prone. Register transfer level is certainly not the only starting point for automated hardware synthesis. The analysis and synthesis of digital circuits are still being pushed toward even higher level of abstraction for manual manipulation. For example, macro block level specification and synthesis has been practiced in some design areas. The register transfer and macro block level design methodologies are part of the architecture and system level techniques to be discussed in this chapter.

A decision made at a higher level of design abstraction generally has more impact on the final quality of the implementation [7.1]. Low power architecture design techniques are important because the design analysis, verification and automated synthesis begin at this level. Typical architecture level design decisions involve the selection and organization of the functional macro of the design. The choice of clocking strategy, parallelism, pipelining, component organization, etc., are issues to be considered at this level.

7.1 Power and Performance Management

Power management refers to a general class of low power techniques that carefully manage the performance and throughput of a system based on its computation needs to achieve low power goals. This is to ensure that we do not waste power by designing hardware that has more performance than necessary. The throughput adaptability of the hardware is also an issue. A good low power system should be able to consume less power when the throughput requirements are low. This class of techniques is typically applied at the architecture level. In the past when power dissipation was not a stringent requirement, the issue of performance and power management is often neglected. In today's complex systems design targeted for low power application, this is a very important feature.

7.1.1 Microprocessor Sleep Modes

The most important and successful use of power management is to deactivate some functional units when no computation is required. For example, the clock supply to a floating point unit can be disabled if not used (see Section 6.1.1). This can be done at different hierarchies and at different levels of the design: subsystems, modules, buses, functional units, state machines, etc. The trade-off is to justify the additional hardware and design complexity in managing the various functional units. At architecture design, this is a particularly attractive technique because little hardware and design complexity is needed to achieve substantial power saving.

In an advanced processor design, power and performance management can be achieved using hardware logic to decide which functional units to be disabled based on the instruction and data stream. This technique is effective when the processor is running at full speed. In another technique, the processor is consciously instructed to reduce its performance by the application software. It may also be halted at a certain state to conserve power while waiting for other events to resume computation.

An excellent example of the power management schemes described above is demonstrated in Motorola's PowerPC 603 processor design [7.2]. The CPU has three primary power saver modes called DOZE, NAP and SLEEP controlled by software. In DOZE mode, most functional units of the processor are stopped except the on-chip cache memory to maintain cache coherency. In NAP mode, the cache is also turned off to conserve power and the processor wakes up after a fixed amount of time or upon an external interrupt. In SLEEP mode, the entire processor clock may be halted and only an external reset or interrupt can resume its operation. The intent is to put more functional units in idle when the processor goes into a deeper sleep mode. However, it requires more latency to resume computation from a deeper sleep mode. This allows the software to choose the proper power saver mode based on the computation requirements of the processor. As computation requirements become less, the processor can be put into deeper sleep modes without any perceivable loss in performance. For example, in a portable computer application, when an application program is running and the processor is waiting for a keystroke, it could be put into the DOZE mode. If the processor is waiting for a disk access, it may go into the NAP mode. If the processor has been idle for a long time and the power saver is activated, the processor can be put into the SLEEP mode.

In PowerPC 603, the sleep modes require less than 10 system clock cycles to resume operation if the system clock and phase-locked-loop (PLL) circuits are not turned off. If the system clock and PLLs are turned off, the chip does not consume any dynamic power but it may take up to 100us to wake up. The various sleep modes of the PowerPC 603 are shown in Table 7.1. The *Dynamic Power Management* mode refers to the application of clock gating at the macro module level in which unused functional units are not clocked.

TABLE 7.1 **Power saver modes of Motorola's PowerPC 603.**

Mode	66Mhz	80Mhz
No power management	2.18W	2.54W
Dynamic power management	1.89W	2.20W
DOZE	307mW	366mW
NAP	113mW	135mW
SLEEP	89mW	105mW
SLEEP without PLL	18mW	19mW
SLEEP without system clock	2mW	2mW

7.1.2 Performance Management

Another interesting performance management technique is reported in an asynchronous data processing system illustrated in Figure 7.1 [7.3]. The system consists of an asynchronous processing unit, a load sensor, a voltage regulator and input/output data buffers. The throughput of the data processing unit is dependent on the operating voltage, which is typical of an asynchronous processing system. The load sensor checks the FIFO buffer length to determine the workload of the system. If the queue is long, the voltage of the system is increased so that the throughput is accelerated. The voltage is scaled down during light loading to conserve power. This adaptive throughput adjustment technique is quite similar to the low power adaptive filtering technique in Section 7.1.3.

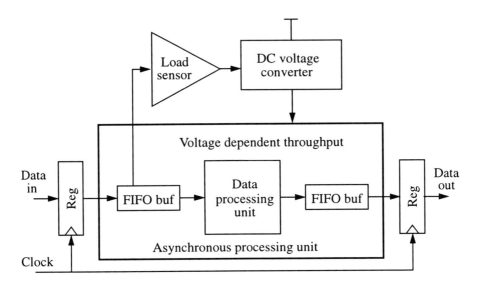

FIGURE 7.1 Adaptive performance management by voltage control.

Performance matching is another important issue in architecture-level management. The performance requirement of a system is typically prescribed from a high level perspective. Since it is a waste of power to design and operate a circuit beyond its performance requirements, it is important to balance the throughput of each subsystem. One form of performance management is to scale down the operating voltage on parts of the system that are not performance critical. The operating voltage of a subsystem can be chosen to match its speed requirement. This allows the performance and power trade-off to be applied at the subsystem level and increases the potential for power efficiency. However, the technique requires the sophistication of multiple voltage supply and logic level converters, which may be a costly feature on a chip. Multiple supply voltage has been used in mixed analog and digital chips, where the analog and digital sections of the chip operate at different voltages. The application of the voltage management technique for low-power digital chips has also been reported [7.4] [7.5].

7.1.3 Adaptive Filtering

One of the first works that exploits the system-level design technique for low power is presented in [7.6]. This system technique is applied to the design of digital filters. The basic principle is to adjust the filter's order length depending on the noise characteristics of the input signal.

A filter is a basic signal processing unit in an electronic system. It allows certain frequency component (pass-band) of the input signal to pass through while rejecting the other signal component (stop-band). For analog signals, filters can be implemented by LC resonance circuits, operational amplifiers, or other electronic devices. A digital filter operates much like an analog filter except that the input and output signals are represented by periodic digital samples of the analog signal. The inputs and outputs are merely streams of numeric values. The digital filtering process is achieved by some arithmetic operations (additions, subtractions and multiplications) of the input samples to produce the output samples.

Two basic classes of digital filter are the *finite impulse response* (FIR) filter and *infinite impulse response* (IIR) filter, as illustrated in Figure 7.2. The output of the filters consists of the sum of products of successively delayed input $x(n)$ and output samples $y(n)$. The delay elements are marked with D in Figure 7.2. The length of the delay chain of the filter is called the *order* of the filter. A higher order filter achieves a higher "quality" of filtering in the sense that the *attenuation* between the pass-band and the stop-band is larger and the *slope* is steeper. This means that the unwanted signal component is better suppressed and the transition region from the pass-band to the stop-band is narrower. Figure 7.3 shows the filter responses of three low pass filters with increasingly higher orders.

The quality or order length requirement of a digital filter depends on:

1. The desired signal-to-noise ratio (SNR) of the output.
2. The noise energy level of the input signal.

If the input signal is less noisy or the desired output SNR is lower, the order length requirement is shorter. In a digital filter design, the output SNR must be above a certain threshold to maintain acceptable signal quality. However, the input SNR often varies with environmental conditions. For example in a voice signal, the input SNR often depends on the speaker, mechanical noise, echo, electromagnetic interference, etc.

For a *fixed order* filter, if the input SNR is higher than normal (i.e., input is not noisy), the output SNR may be much higher than that required by the specification. This means that a lower order filtering should be sufficient to achieve the specified output

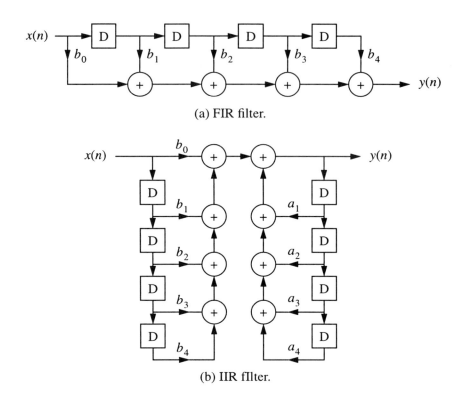

(a) FIR filter.

(b) IIR fIlter.

FIGURE 7.2 Digital filter architectures.

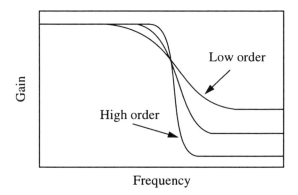

FIGURE 7.3 Filter responses with increasing filter order.

SNR. As evident from the filter architecture in Figure 7.2, a lower order filter requires less power to operate because the amount of computation is less. From this observation, a low power technique to adaptively adjust the filter order based on the input SNR has been proposed [7.6]. When the input signal is less noisy, the filter order is reduced to conserve power. When the input noise increases, the filter order is lengthened accordingly. The output SNR of the filter will be maintained above the minimum threshold specified by the design thus achieving the desired output signal quality.

To adjust the filter order, we need to estimate the output SNR or equivalently the noise energy of the filter's output. This has been a well-studied subject in adaptive filtering. The filter input consists of a signal component in the pass-band and a noise component in the stop-band. The output signal, with much of the noise component filtered, is a good estimate of the true signal component. Therefore, the difference between the input energy and the output energy of the filter is a very good estimate of the input noise energy. The energy difference is given by

$$d(n) = \frac{1}{L} \left[\sum_{k=0}^{L-1} |x(n-k)|^2 - \sum_{k=0}^{L-1} |y(n-k)|^2 \right] \qquad (7.1)$$

where L is the filter length, $x(n)$ and $y(n)$ are the input and output samples of the filter. The observed difference $d(n)$ is the input signal energy being filtered, which is a good estimate of the noise energy. The general strategy of adaptive filtering is to adjust the filter order based on the noise energy estimates. A general framework for adjusting the filter order based on the observed noise energy is shown in Figure 7.4. Various adaptive filter order update schemes based on the noise estimate can be applied to provide the trade-off in filter quality, power budget and design complexity.

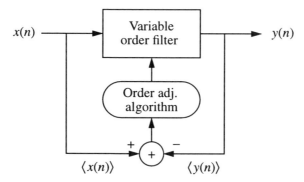

FIGURE 7.4 A general framework for updating filter order length.

173

We will describe an adjustment algorithm proposed in [7.6]. The filter order is evaluated at each data sample and the order is adjusted if necessary. The actual noise energy of the output can be estimated by

$$Q(n) = \alpha\, d(n)\, E_{SB}[Order(n)] \qquad (7.2)$$

where $Order(n)$ is the filter order (which varies with time); $E_{SB}[Order(n)]$ is the stop-band energy of the filter and α is a constant. The energy $E_{SB}[Order(n)]$ can be obtained by taking the integration of the filter transfer function at the stop-band and stored in a lookup table. $Q(n)$ represents the noise energy measured at a particular output sample n. The filter order update algorithm is

$$Order(n) = \begin{cases} Order(n-1)+N, & Q(n) > MAX \\ Order(n-1), & MIN \le Q(n) \le MAX \\ Order(n-1)-N, & Q(n) \le MIN \end{cases} \qquad (7.3)$$

The variable N is a constant in which the filter order is adjusted at each sample. When the output noise energy $Q(n)$ rises above the MAX threshold, the filter order is increased by N to improve the filter quality. Conversely, when $Q(n)$ drops below MIN, the filter order is decreased by N to conserve power. It can be shown that computing $Q(n)$ requires 5 multiplications and 5 additions. This is the overhead of implementing the low power technique, roughly equivalent adding 5 orders to an FIR filter.

The net benefit of this low power technique depends on the noise characteristics of the input signal. If the filter input is constantly in a very noisy state that maximizes the filter order, having the adaptive mechanism may lead to more power being consumed. Power conservation is only achieved when the input noise is low most of the time, with occasional surges in noise level. In the above adaptive algorithm, the overhead represents an additional five-order FIR filter (or a fixed order depending on the adaptive mechanism). If the high-noise and low-noise states of the input signal do not correspond to filter order difference of more than five, no power conservation will be achieved. Such dramatic noise variation characteristics may not always present in a signal processing system.

In [7.6], simulation experiments have been performed to evaluate the potential power saving of the low power technique. It was observed that up to 10X power reduction is possible compared to a straightforward fixed order filter design in speech signal processing. As compared to other low power techniques at a lower abstraction level, which have been discussed in the previous chapters, the power reduction of such scale is phenomenal. However, this technique is only applicable to the design of digital filters. It exploits a particular nature of the system, i.e., variation of input SNR, to achieve the power efficiency. This is a manifestation that higher level techniques tend to be more specialized but provide greater payback.

7.2 Switching Activity Reduction

Switching activities are the biggest cause of power dissipation in most CMOS digital systems. In order to do computation, switching activities cannot be avoided. However, some switching activities do not contribute to the actual computation and should be eliminated. Such switching activities are results of undefined or unspecified behavior of a system. For example, when the result of a multiplier is not used, the inputs of the multiplier are arbitrary and they will not affect the functional correctness of the system. However, if the multiplier inputs are allowed to switch, unnecessary switching activities will be introduced.

The suppression of switching activities always involves some trade-off decisions. In general, hardware logic is required to suppress unwanted switching activities and the additional logic itself consumes power. Therefore, it is important to estimate how many switching activities can be eliminated by a particular technique so that it can be justified.

7.2.1 Guarded Evaluation

Guarded evaluation [7.7] is a technique to reduce switching activities by adding latches or blocking gates at the inputs of a combinational module if the outputs are not used. An example is depicted in Figure 7.5 in which the result of the multiplication may or may not be used depending on the condition selection of the multiplexor. To reduce switching activities, latches are added at the inputs of the multiplier. The latches are transparent when the result of the multiplication is to be used. Otherwise, the latches preserve the previous values of the multiplier inputs to suppress activities inside the multiplier because the result will not be used. It is a special case of the pre-computation technique introduced in Section 5.5.

Instead of using latches, which increase the area substantially, we can also use AND-gates to mask one or both inputs of the multiplier to zero. This forced zero method still introduces some useless activities in the multiplier when the selection condition changes. But if the zero-out condition changes infrequently compared to the data rate, this method will achieve a similar effect with less hardware added.

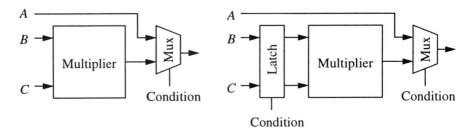

FIGURE 7.5 Guarded evaluation.

7.2.2 Bus Multiplexing

One common way to reduce hardware resources in a digital system is to share long data busses with time multiplexing. This scheme is shown in Figure 7.6(b). At even clock cycles, source S_1 uses the shared bus to send data to destination D_1 while at odd cycles, source S_2 sends data to D_2.

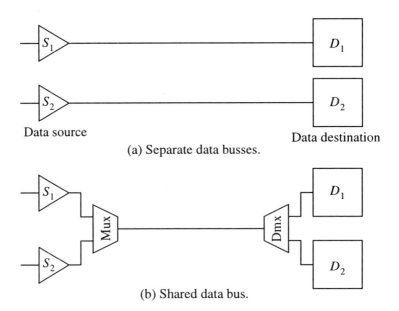

FIGURE 7.6 Data bus sharing with time multiplexing.

In digital signal processing systems, a phenomenon that is advantageous to power efficiency has been observed: correlation of data samples. Correlation refers to the fact that successive data samples are quite similar in values and consequently their binary representations have many bits in common. This is a direct result of sampling a band-limited analog signal with a relatively high sampling rate. Methods of analyzing data correlation have been discussed in Section 2.5.

In two's complement representation, the sign bit section of data bus signals rarely switches when the data stream is positively correlated. This is desirable for power efficiency. However, if the data bus is shared, the correlation is often lost because the shared bus carries data samples from two uncorrelated data sources. The switching activity characteristics of the shared and dedicated busses are shown in Figure 7.7.

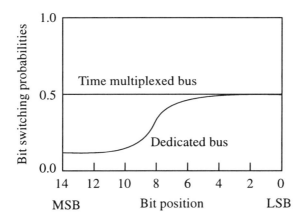

FIGURE 7.7 Switching activities of a positively correlated data stream.

Thus to improve power efficiency, bus sharing should not be used for a positively correlated data stream. In a *negatively correlated* data stream (see Section 2.5), successive samples often switch the sign bits. So, sharing a bus with another uncorrelated data stream can *improve* the power efficiency because the shared bus tends to have randomly switching characteristics.

7.2.3 Glitch Reduction by Pipelining

Glitches are unwanted switching activities that occur before a signal settles to its intended value. They are caused by transitions at multiple inputs of a gate within a clock cycle. Glitches are undefined and unpredictable behavior of a circuit that cannot be analyzed deterministically. Glitches of a signal node are dependent on the *logic depth* of the node, i.e., the number of logic gates from the node to a primary input or a sequential element. Generally, nodes that are logically deeper are more prone to signal glitches. One of the reasons is that the transition arrival times of the gate inputs are spreaded longer due to delay imbalances. Also, a logically deep node is typically affected by more primary input switching and therefore more susceptible to glitches. One way to reduce glitches is to shorten the depth of combinational logic by adding pipeline registers. This is very effective especially for datapath components such as multipliers and parity trees. Datapath components are more glitch-prone because they have deep logic and the inputs switch simultaneously.

7.3 Parallel Architecture with Voltage Reduction

Parallelism has been traditionally used to improve the computation throughput of high performance digital systems. It has been successfully used in advanced microprocessors operating at the frequency limited by the process technology. Compared to a uniprocessing system, parallelism effectively increases the throughput of a system by performing simultaneous computations without increasing the operating frequency. Since operating frequency and throughput is proportional, we can view parallelism from another perspective: it *lowers* the operating frequency while maintaining the same system throughput. Parallelism essentially trades area off for a lower operating frequency or higher throughput. The same trade-off idea can also be applied to achieve power reduction.

When the system design of a chip has the luxury of defining the system operating voltage, an interesting low power technique using parallel processing can be used [7.1] [7.8]. This stems from the observation that a higher voltage system can be operated at a higher frequency because of shorter average gate propagation delay. In Equation (4.13), we showed that the maximum operating frequency of a CMOS gate is inversely proportional to its operating voltage. As the operating voltage approaches the transistor threshold voltage, the gate delay increases, thus effectively limits the maximum operating frequency of the system. However, a low voltage system is attractive for the obvious benefit of low power dissipation. To overcome this problem, parallelism can be applied so that the frequency requirement of the system is reduced, thus allowing the system designer to choose a lower operating voltage.

Consider a digital signal processing system in which a throughput requirement corresponding to a frequency of f has to be satisfied. In a uniprocessing system, the power dissipation will be given by

$$P_{uni} = CV^2 f \qquad (7.4)$$

where C is the average capacitance switched and V is the operating voltage the uni-processing system.

If we double the number of processing units, each unit can be operated at half the frequency $f/2$. The desired system throughput is still maintained by multiplexing the outputs of the two processing units as shown in Figure 7.8. Since the operating frequency requirement is now reduced to half, the system operating voltage can be reduced to conserve power. As a thought experiment, let's assume that the average capacitance switched in the parallel system is $2.2C$, slightly more than double due to some overhead in multiplexing and demultiplexing the data. Suppose that the reduced operating frequency allows us to reduce the operating voltage of the parallel system to $0.6V$. The power dissipation parallel system is therefore

$$P_{par} = (2.2C)(0.6V)^2(0.5f) = 0.396 P_{uni} \qquad (7.5)$$

representing a power reduction of approximately 60% compared to the uniprocessing system. The power saving is paid by a substantial increase in circuit area.

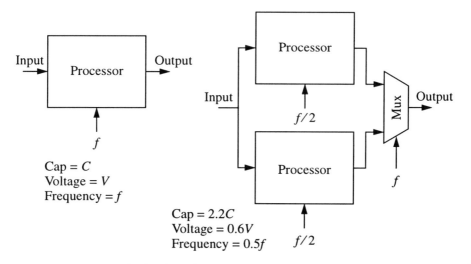

FIGURE 7.8 Power dissipation of uniprocessing and parallel processing systems.

The above parallel technique increases the chip area by at least twice. If the area penalty of a parallel system is prohibitive, *pipelining* can offer similar trade-off results with less area overhead but more complexity in controller design. Pipelining, as

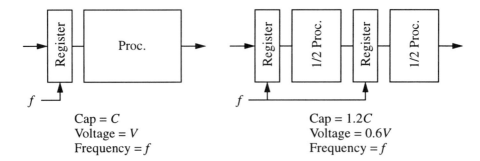

FIGURE 7.9 Power efficiency of a pipelined system.

depicted in Figure 7.9, is another form of parallelism. A pipeline register is added to the original uniprocessing design. Since the propagation delay of the processing unit is reduced to half, we can operate the pipelined system with a lower voltage while maintaining an identical overall throughput. Suppose the pipelining overhead increases the capacitance to $1.2C$ while reduces the voltage to $0.6V$, the power dissipation of the pipelined system is

$$P_{pip} = (1.2C)(0.6V)^2 f = 0.432 P_{uni} \tag{7.6}$$

In theory, we can parallelize or pipeline a uniprocessing system n times to improve its power efficiency. Let us derive the theoretical power efficiency of such a system. Assume that area and voltage scale linearly with the parallelism factor. If we parallelize the uniprocessing system n times, the hardware is duplicated n times resulting in nC capacitance. To maintain the overall throughput, each piece of hardware will only require $1/n$ system throughput resulting in operating frequency of f/n. The delay of a circuit is assumed to be proportional to the voltage supply and therefore each unit can be operated at voltage V/n. The theoretical limit of the power dissipation of the parallel system is

$$P_{par} = (nC)\left(\frac{V}{n}\right)^2\left(\frac{f}{n}\right) = \frac{P_{uni}}{n^2} \tag{7.7}$$

In an n-stage pipelined system, the delay requirement of each stage is reduced by a factor of n. Hence, the voltage can be reduced to V/n while the operating frequency f remains unchanged. The total capacitance of the hardware remains the same since there is no hardware duplication (ignoring pipeline registers and control circuitry). The power dissipation of the pipelined system is

$$P_{pip} = C\left(\frac{V}{n}\right)^2 f = \frac{P_{uni}}{n^2} \tag{7.8}$$

Of cause the idealized analysis ignores the overhead in voltage, area and frequency scaling. The parallel implementation increases the area by a factor of n and the overhead in data multiplexing/demultiplexing and controller circuitry was ignored. In the pipelined system, the complexity of the system operation increases significantly. The increase in clock loading, additional management circuitry, pipeline latency and the degradation caused by unbalanced delay distribution on the pipeline have been ignored.

Although the general principle of parallelism is applicable to any digital systems, this low power technique is more appropriate for special purpose architecture digital signal processing systems that have intensive data operations and relatively simple control structures. Duplicating or pipelining data operators in such systems are easier compared to general purpose processors with complicated control structures.

There are some difficulties in the application of this technique. In most system design, the operating voltage may have been dictated by other factors beyond the designer's control. Sometimes the system operating voltage is already at the lowest offered by processing technology. Other times, the availability of low voltage circuits or cell libraries may be limited or require some costly redesign efforts.

7.4 Flow Graph Transformation

In this section, we focus on a system level technique for the design of special purpose architecture digital signal processing systems. Such systems are characterized by computation intensive data path operations with simple control structures. The system architecture can be represented by a *control data flow graph*. The graph consists of *control nodes* and *data nodes* connected by *directed edges*. Control nodes change the flow of data that pass through it. Examples of control nodes are multiplexors, condition selectors, branch and delay elements, etc. Data nodes provide computation operators for the input data streams such as addition, multiplication, shift, etc. The graph edges represent the data streams of the system.

A control data flow graph expresses the conceptual computation algorithm of the system. It provides basic information such as the number of control and data operators, their ordering dependencies and the inherent parallelism that may exist. The path from system input to output is a good estimate of the delay of the system. The number of operator nodes is an estimate of the computation needs and the complexity of the system.

A control data flow graph is often the starting point to derive the actual hardware architecture of a system by mapping the operators and edges to actual hardware modules and busses respectively. A controller schedules the operators to perform the desired computation in the proper order. Figure 7.10 illustrates a system that computes the equation $y_n = a_n b_n + 3a_{n-1}$ and its corresponding control data flow graph. The hardware architecture that executes the computation is shown in Figure 7.10(b). To reduce area, we decided to use only one multiplier module with a two-phase scheduling for each data sample (resource sharing). In phase one (T_0), the multiplier computes $a_n b_n$ and stores the result in the lower register. In phase two, it computes $3a_{n-1}$ and stores the result in the upper register. In the next data cycle, the partial terms are added and stored in the output register. The result of addition could be scheduled at either T_0 or T_1 and we arbitrarily chose T_0. There is one data sample latency between the input and output, i.e., the output register contains the result of the previous input samples.

The above example illustrates a simple mapping from the control data flow graph to the system hardware architecture. The hardware realization requires additional design details such as control step scheduling, resource sharing, register allocation, etc. It should be noted that many different mappings to system hardware are possible. For the same graph, a different mapping may alter the power, area and delay characteristics of the hardware architecture. Nevertheless, there is enough information at the control data flow graph level to perform power analysis, optimization and trade-off, as will be discussed in the following sections.

7.4.1 Operator Reduction

We now examine some operator transformations on a control data flow graph. The transformations preserve the functionality of the graph so that the resulting graph is computationally equivalent to the original graph. The transformed graph represents an alternate hardware implementation with a different trade-off. If the system operating voltage is variable, the transformed graph may offer a lower voltage implementation by reducing the worst case delay path [7.9].

An example of operator transformation is illustrated in Figure 7.11 [7.10]. The multiplication by an integer constant can be replaced by binary shift (multiplication by powers of 2) and add/subtract operations, which may be more power efficient.

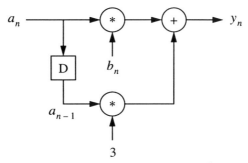

(a) Control data flow graph of $y_n = a_n b_n + 3a_{n-1}$.

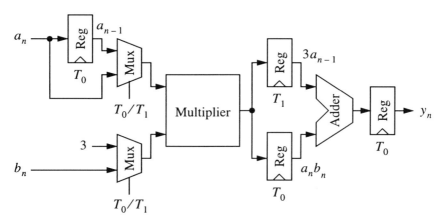

(b) Hardware architecture and scheduling.

FIGURE 7.10 Control data flow graph and its mapping to hardware architecture.

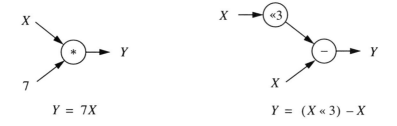

$$Y = 7X \qquad\qquad Y = (X \ll 3) - X$$

FIGURE 7.11 Constant multiplication versus shift and addition.

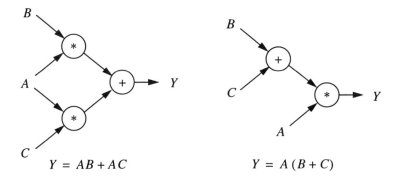

$$Y = AB + AC \qquad\qquad Y = A(B + C)$$

FIGURE 7.12 Associative transformation.

Rearranging the order of computation can also lead to fewer operations and lower power, as shown in Figure 7.12. The familiar coordinate rotation equations with a constant angle θ

$$\begin{bmatrix} x_2 \\ y_2 \end{bmatrix} = \begin{bmatrix} \sin\theta & \cos\theta \\ \cos\theta & -\sin\theta \end{bmatrix} \begin{bmatrix} x_1 \\ y_1 \end{bmatrix} \tag{7.9}$$

can be rewritten as

$$x_2 = x_1 \sin\theta + y_1 \cos\theta = (x_1 + y_1) \sin\theta + y_1 (\cos\theta - \sin\theta)$$
$$y_2 = x_1 (1 - \sin\theta) - y_1 \sin\theta = x_1 - (x_1 + y_1) \sin\theta \tag{7.10}$$

and implemented with only two multiplications and three additions/subtractions. Direct and transformed implementations are shown in Figure 7.13. The term $\cos\theta - \sin\theta$ is a precomputed constant independent of the inputs x_1, y_1 and does not require an add operator. Note that the direct implementation based on Equation (7.9) is not necessary inferior in power dissipation. The direct implementation has a shorter critical delay path and therefore could operate at a lower voltage, as explained by the parallelism technique in Section 7.3.

The above transformations exploit the inherent efficiency that exists in the arithmetic computations. Such transformation can often be identified from the computation structures of the system. The transformations change the power, area and delay trade-off of the system. Some transformations reduce power, area and sometimes delay paths.

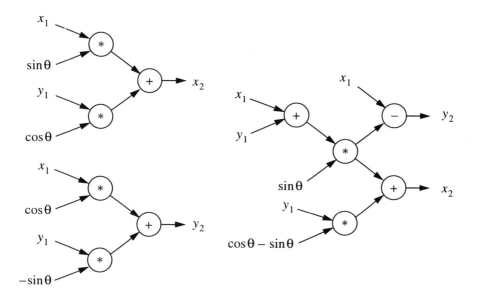

FIGURE 7.13 Direct and transformed flow graphs of coordinate rotation operation.

7.4.2 Loop Unrolling

The control data flow graphs of digital signal processing systems often contain loops, as a result of recursive computations. An important technique for flow graph transformation is to *unroll* the loop [7.9]. Loop unrolling is a method to apply parallelism to the computation. Consider a simple recursive computation of an IIR filter

$$y_n = b_0 x_n + a_1 y_{n-1} \qquad (7.11)$$

We can unroll the loop once and obtain

$$y_n = b_0 x_n + a_1 b_0 x_{n-1} + a_1^2 y_{n-2}$$
$$y_{n-1} = b_0 x_{n-1} + a_1 y_{n-2} \qquad (7.12)$$

The data flow graphs of the original and unrolled computation are depicted in Figure 7.14.

After loop unrolling, two output values are produced using two input values in a single computation cycle. The unrolled computation structure increases the computation by more than twice. The number of multiplications increases from 2 to 5 and the num-

185

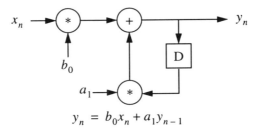

$$y_n = b_0 x_n + a_1 y_{n-1}$$

(a) Original control data flow graphs.

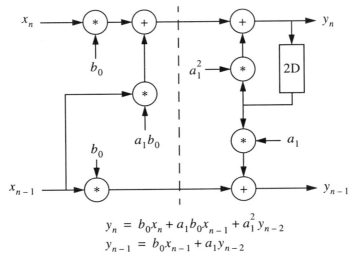

$$y_n = b_0 x_n + a_1 b_0 x_{n-1} + a_1^2 y_{n-2}$$
$$y_{n-1} = b_0 x_{n-1} + a_1 y_{n-2}$$

(b) Unrolled control data flow graph.

FIGURE 7.14 Loop unrolling transformation.

ber of additions increases from 1 to 3. If the unrolled computation structure is implemented directly, the power efficiency should be worst than the original implementation because of the increased computation. However, the unrolled structure allows us to apply pipelining by adding pipeline registers on the graph edges crossing the vertical dashed line in Figure 7.14(b). With pipelining, the longest delay path of the unrolled graph is identical to the original graph in Figure 7.14(a), i.e., a multiplication followed by an addition. Since the unrolled graph produces two outputs simultaneously, it can be implemented with half the operating frequency of the original graph. Thus, its critical delay is identical but the operating frequency is halved. This allows us to lower the operating voltage to improve the overall system power efficiency.

In summary, loop unrolling is a method to apply parallelism to a recursive computation structure. It can be considered a special case of the parallelism with voltage reduction technique described in Section 7.3. Loop unrolling itself does not result in more efficient computation. It even incurs a penalty because the amount of computation more than doubles that of the original graph. However, the unrolled graph enables other low power techniques such as operator transformation, pipelining and parallelism, which are not possible with the original flow graph. The power efficiency can also be gained by lowering the system operating voltage.

REFERENCES

[7.1]   A. Chandrakasan, S. Sheng and R. Brodersen, "Low-power CMOS Digital Design," *IEEE Journal of Solid-State Circuits*, vol. 27, no. 4, pp. 473-484, Apr. 1992.

[7.2]   G. Gerosa, *et al.*, "A 2.2 Watt 80 MHz Superscalar RISC Microprocessor," *IEEE Journal of Solid-State Circuits*, vol. 29, no. 12, pp. 1440-1454, Dec. 1996.

[7.3]   L. Nielsen, C. Niessen, J. Sparso and K. van Berkel, "Low-Power Operation Using Self-Timed Circuits and Adaptive Scaling of the Supply Voltage," *IEEE Transactions on VLSI Systems*, vol. 2, no. 4, pp. 391-397, Dec. 1994.

[7.4]   K. Usami and M. Horowitz, "Clustered Voltage Scaling Technique for Low-Power Design," *Proceedings of International Symposium on Low Power Design*, pp. 3-8, 1995.

[7.5]   K. Usami, *et al.*, "Automated Low-power Technique Exploiting Multiple Supply Voltages Applied to a Media Processor," *Proceedings of Custom Integrated Circuit Conference*, pp. 131-134, 1997.

[7.6]   J. Ludwig, S. Nawab and A. Chandrakasan, "Low-Power Digital Filtering Using Approximate Processing," *IEEE Journal of Solid-State Circuits*, vol. 31, no. 3, pp. 395-400, Mar. 1996.

[7.7]   V. Tiwari, S. Malik and P. Ashar, "Guarded Evaluation: Pushing Power Management to Logic Synthesis/Design," *Proceedings of International Symposium on Low Power Design*, pp. 221-226, 1995.

[7.8]   A. Chandrakasan, M. Potkonjak, J. Rabaey and R. Brodersen, "HYPER-LP: A system for Power Minimization using Architectural Transformations," *Proceedings of International Conference of Computer-Aided Design*, pp. 300-303, 1992.

[7.9]   A. Chandrakasan, M. Potkonjak, R. Mehra, J. Rabey and R. Brodersen, "Optimizing Power Using Transformations," *IEEE Transactions on Computer Aided Design*, vol. 14, no. 1, pp. 12-31, Jan. 1995.

[7.10]   M. Potkonjak, M. Srivastava and A. Chandrakasan, "Efficient Substitution of Multiple Constant Multiplications by Shifts and Additions using Iterative Pairwise Matching," *Proceedings of Design Automation Conference*, pp. 189-194, 1996.

CHAPTER 8

Advanced Techniques

The technology landscape of VLSI design and fabrication is constantly changing. The techniques for low power design will undoubtedly be affected by the progress in processing technology and design methodology. As we look beyond the current technology profile, some promising low power design techniques appear on the horizon. These design techniques, such as pass transistor logic and asynchronous logic, are not overnight revolutionary breakthroughs. Rather, they are techniques practiced by a few designers in some very narrow application areas. We will examine the characteristics of these design and logic styles, their efficiencies in power, area, performance and other design merit criteria.

None of the techniques to be discussed has been recognized in the main stream digital VLSI design domain. It is not assured that any of the techniques will dominate in future. Some techniques are still very labor-intensive and there is a lack of design automation software that can exploit the low power techniques. We will discuss the advantages and limitations of each technique with the perspective of our current technology. It should be noted that as technology progresses, some of the advantages may be diminished. On the other hand, some of the limitations may be relaxed in future.

To date, these advanced low power techniques are very controversial. Most arguments, whether approve or disapprove, have been mostly based on predictions and philosophical preferences rather than empirical scientific observations. One reason is the lack of technical discourse and publication with repeatable experiments in the engineering community. Thus, it is not the intent of the author to promote or discourage a particular technique. This chapter merely relates the known phenomena and facts of the design techniques and reserves the judgement to the readers.

8.1 Adiabatic Computation

In Section 1.2, we derived the fundamental dynamic power dissipation equation $P = C_L V^2 f$ and stated the conditions in which the equation is applied. The conditions are:

1. Capacitance C_L is constant.
2. Voltage V is constant.
3. The capacitor is fully charged and discharged.

We also note that almost all CMOS digital integrated circuits today satisfy the above conditions. Throughout this book, all low power techniques studied so far have not been able to escape the confinement of the equation. We merely tried to reduce the contribution of one or more variables of the equation to conserve power.

In this section, we seek out to violate one of the conditions above and, for the first time, will construct logic circuits that can surpass the limitation of the $P = C_L V^2 f$ power equation. We will no longer restrict ourselves to circuits where the supply voltage is a constant. We introduce a logic family in which the supply voltage is varied with the logical operations. It resulted in a logic computation with energy efficiency better than $C_L V^2$ per signal transition.

The circuit technique is commonly referred as *adiabatic* because in an ideal situation, it does not consume power. However, as we will see, the zero power condition can never be achieved in any practical computation system. More appropriately, the technique is called "asymptotically zero power" because the dynamic energy dissipation per computation cycle approaches zero as the system computation frequency approaches zero. On the contrary, the dynamic energy dissipation per computation cycle of the conventional CMOS circuit remains a constant $C_L V^2$ independent of the system frequency. Comparison of adiabatic and static CMOS logic is detailed in [8.1].

8.1.1 Complementary Adiabatic Logic

We introduce a type of adiabatic logic originally proposed in [8.2] [8.3]. The circuit uses fully differential logic signals and requires a four-phase dynamic operation to compute a single Boolean logic function. Figure 8.1 shows the circuit of an inverter.

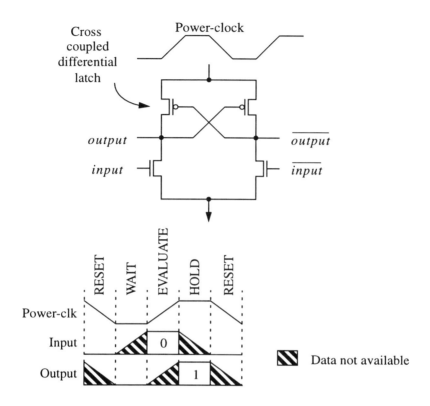

FIGURE 8.1 A four phase adiabatic logic inverter.

There are several noticeable differences between the circuit and a conventional static or dynamic CMOS circuit. The most evident difference is that the supply voltage is no longer constant. A periodic ramp signal is used to power the circuit. The power signal also acts as the clock signal of the circuit because it synchronizes the circuit operations.

Each logic bit requires two complementary signals for inputs and outputs. The P-transistor pair provides positive feedback to the output signals, like a cross coupled inverter. The two N-transistors of the inverter acts as evaluation transistors during logic computation, much like the dynamic CMOS logic. In general, the N-transistors form a serial-parallel network to perform a non trivial Boolean computation. The circuit requires four phase operations to achieve a Boolean computation step.

The four phases of operation are described below:

1. At the RESET phase, both inputs are held at the ground level by the previous logic stage, which is at the WAIT phase. This essentially shuts off all N-transistors so that when power is ramping down, the outputs lose their differential signals from the previous computation cycle.

2. During the WAIT phase, the inputs start to develop differential voltage because the preceding gate is at the EVALUATE phase. In this phase, the two output signals are held at the ground level, as this is required to sustain the proper operation of the succeeding gate.

3. At the end of the WAIT phase, the input differential voltages have fully developed, and the EVALUATE phase begins. During this phase, power ramps up and the input differential voltage is sustained by the preceding gate. The input differential causes one of the N-transistors to turn on, clamping one of the output signals to the ground level. As power ramps up, the other output that is not tied to ground will also ramp up because of the cross coupled P-transistors. At the end of the EVALUATE phase, the output differential voltage is fully developed to drive the succeeding gate.

4. During the HOLD phase, the P-transistors maintain the output differential voltage. In this cycle, the preceding gate is at the RESET phase and the input signals lose their differential voltages. At the end of the HOLD phase, the inverter returns to the RESET phase and completes the four-phase computation cycle. The next computation cycle will start in the next phase.

Boolean logic computation is achieved from the EVALUATE phase to the HOLD phase. The outputs at the HOLD phase compute the logic function using the input signals received during the EVALUATE phase. The other phases are required for the proper lock-stepped operation of the adiabatic gate.

Note that during the HOLD phase when the output is valid, one of the outputs tied to the ground level becomes floating near the end of the HOLD phase. When one of the input signals ramps down below the N-transistor threshold voltage, the grounded output is essentially floating with both P- and N-transistors turned off. The circuit relies on the charge storage effect of the node for proper operation. This places a lower bound on the operating frequency of the circuit and weakens the noise immunity. A proposed solution is to use two fully complementary inverters to replace the cross coupled P-transistors, much like a 6-transistor SRAM cell. The circuit is shown in Figure 8.2. The operation of the circuit is identical to that of the original circuit.

8.1.2 Power Efficiency of Adiabatic Logic

How does the complementary adiabatic logic computation differ from the static CMOS gate in power dissipation? The equivalent circuit for the conventional

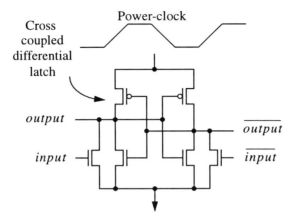

FIGURE 8.2 Adiabatic circuit with full complementary feedback.

dynamic or static CMOS logic is shown in Figure 8.3(a). Recall Section 1.2 where the $P = C_L V^2 f$ equation is derived, the energy E_s dissipated during a charging and discharging cycle according Equations (1.5) and (1.6) is

$$E^s = E_c + E_d = C_L V^2 \tag{8.1}$$

regardless of the resistor R_s. E_c (E_d) is the energy dissipated due to charging (discharging) of the capacitor C_L.

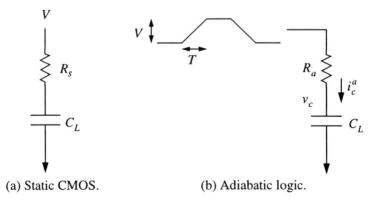

    (a) Static CMOS.   (b) Adiabatic logic.

FIGURE 8.3 Equivalent circuit when charging output capacitance.

The charging (discharging) of parasitic output capacitance in the adiabatic logic circuit occurs in the EVALUATE (RESET) phase. The equivalent circuit for the charging phase is shown in Figure 8.3(b) where R_a is the resistance of the charging P-transistor. We assume that the time constant of the circuit $R_a C_L$ is much smaller than the voltage ramp period T, i.e.,

$$R_a C_L \ll T \tag{8.2}$$

This means that the voltage at the capacitor closely follows the supply voltage and the potential difference across R_a is very close to zero. The voltage at the capacitor v_c is also a constant ramp with slope V/T where V is the swing of the supply voltage and T is the power-clock phase period. The charging current of the adiabatic circuit is

$$i_c^a = C_L \frac{dv_c}{dt} = \frac{C_L V}{T} \tag{8.3}$$

and the energy dissipated in the resistor is

$$E_c^a = (i_c^a)^2 R_a T = \left(\frac{C_L V}{T}\right)^2 R_a T = \left(\frac{R_a C_L}{T}\right) C_L V^2 \tag{8.4}$$

The equivalent circuit for discharging is similar to that of charging except that the supply voltage ramps down. Therefore, Equations (8.3) and (8.4) also apply to the discharging phase and the energy dissipation of a four-phase computation cycle of the adiabatic logic is

$$E^a = \left(\frac{2R_a C_L}{T}\right) C_L V^2 \tag{8.5}$$

The difference between Equations (8.1) and (8.5) is the factor $(2R_a C_L)/T$. In both cases, the electrical energy from the power supply is dissipated by the resistance of the charging and discharging circuit only. Capacitors only act as energy storage devices and do not dissipate power. In CMOS circuits where the supply voltage is constant, a large voltage is exerted across the resistor R_s when the charging cycle begins. This results in large current and energy dissipation. However, in the adiabatic logic, a slow ramping power signal is used to charge the capacitor. The voltage across the resistor R_a is much smaller compared to that of the static CMOS circuits, resulting in less energy dissipation. Consider the energy stored in the capacitor $0.5 C_L V^2$ and the energy dissipated during charging or discharging E_c^a. Because $R_a C_L \ll T$, we have

$$E_c^a = \left(\frac{R_a C_L}{T}\right) C_L V^2 \ll \frac{1}{2} C_L V^2 \tag{8.6}$$

The energy stored in the capacitor is more than that dissipated by the resistor during charging or discharging. This suggests that during discharging, some energy stored in the capacitor is actually *returned* to the energy source. In the adiabatic logic circuit, the power supply is implemented by a resonant circuit that can reclaim the electrical energy stored in the capacitor and reuse it in the next cycle. The resonant circuit is crucial to achieve the desired power efficiency of the adiabatic logic. Otherwise, the energy may be dissipated as heat in the power supply circuitry.

The efficiency of the adiabatic logic compared to that of the static CMOS logic is reflected in the term $(2R_a C_L)/T$ in Equation (8.5). The resistance R_a is primarily contributed by the on-resistance of the P-transistor in the circuit. The factor $2R_a C_L$ is a constant determined by the process technology, transistor sizes and the parasitic capacitance. The variable T is the phase period of the computation specified by the system performance. In a typical bulk CMOS technology today, $2R_a C_L$ is in the order of 0.1ns to 1ns. With a system operating frequency of 100MHz, T is at 10ns. This represents one to two orders of magnitude improvement in power efficiency over the static CMOS logic. If the system clock frequency is lowered, the efficiency increases making adiabatic logic very attractive. Thus, the adiabatic logic style is probably more suitable for low performance, low power applications. When the period T approaches $2R_a C_L$, the efficiency of adiabatic logic diminishes because the potential difference across R_a increases. Figure 8.4 illustrates the efficiency of the adiabatic and static CMOS logic. The energy efficiency of static CMOS is independent of the operating frequency. However at low operating frequencies, adiabatic circuits are more energy efficient than static CMOS circuits. As the frequency increases, energy efficiency decreases and eventually the overhead in the adiabatic logic becomes intolerable.

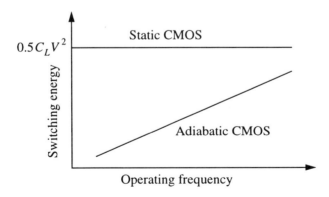

FIGURE 8.4 Switching energy of static CMOS and adiabatic logic.

195

From the practical standpoint, there are several limitations of the complementary adiabatic logic. First, the transistor count per logic gate is more than that of the static CMOS. The four-phase operation introduces additional complexity in the design process. Note that because of the differential output signals, one of the output capacitors is always charged and discharged regardless of the output logic value. This means that each gate will *always* switch when performing a logic computation. In static CMOS logic, the output switches only when it is different from its previous value. Some adiabatic circuits use single wire logic and avoid this problem.

In practice, the power-clock signal is generated by a resonant circuit with sine wave output to approximate the ramp signal. The resonant circuit itself consumes some power and typically requires inductors. This introduces additional complexity in the system design. For example, it is difficult to design "sleep modes" with the clock signal muted because the resonant circuit has to be stopped and restarted.

As indicated in Equation (8.5), the efficiency depends on the ratio of $R_a C_L$ and T. As process technology improves, the ratio becomes smaller if T is to remain the same. The practical value of adiabatic logic as a mainstream logic computation style is still questionable. It is generally agreed that adiabatic logic offers better power efficiency at low frequency range. There have been conflicting viewpoints about the merits of the adiabatic circuit techniques at higher frequency [8.4] [8.5] [8.6]. The operations of adiabatic principle have been proven possible for logic circuits [8.3] and SRAMs [8.7] in prototype designs. Large scale production application of the circuit has not been reported.

8.2 Pass Transistor Logic Synthesis

Existing logic synthesis systems are based on very limited types of logic cells. These systems are suitable for handling computation expressed in the form of Boolean equations using AND/OR/INVERT operators. Such functions can be readily mapped to CMOS cells such as NANDs, NORs and AOIs. AND/OR/INVERT operators are natural to human specification and manipulation. However in a logic synthesis system, the Boolean functions used are far more complex than the simple ANDs and ORs. They include multiplexor functions, majority functions, exclusive-OR functions, etc. If the complex functions are available as cells, the logic synthesis system can make use of them without constructing the functions from the basic AND/OR/INVERT gates. Pass transistor logic can implement certain complex Boolean function efficiently. Therefore, there may be certain merits for applying pass transistor logic in complicated Boolean computation systems.

In the past, pass transistor logic has been used in handcrafted full custom design. This somewhat limits the application of the circuit technique because more circuits are now synthesized rather than individually handcrafted to increase design productivity. The main reason for the limited use of pass transistor logic is the lack of design automation tools for synthesizing pass transistors. In this section, we will discuss the merits and demerits of pass transistor logic. We will briefly describe a pass transistor logic synthesis system [8.8] which has shown to have better power and area efficiency than the conventional CMOS style synthesis system.

8.2.1 Basics of Pass Transistor Logic

Pass transistor logic uses *pass transistors* to compose Boolean logic functions. A pass transistor is a MOS device that acts like a switch. It "passes" logic value from input to output when the device is turned on and isolates the input from the output when the device is turned off. An example of pass transistor logic implementing a two-input multiplexor is depicted in Figure 8.5. The select input turns on one of the pass transistors. This allows the input A or B to propagate to the output Y depending on the logic value of the select signal.

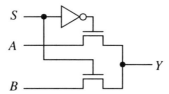

FIGURE 8.5 A two-input multiplexor using pass transistor logic.

An NMOS pass transistor allows the logic 0 signal to pass through without any signal degradation. However, if a logic 1 (voltage level V_{dd}) is exerted at the input of an NMOS pass transistor, the output can only rise to $V_{dd} - V_{tn}$ due to the threshold voltage V_{tn} of the N-transistor. We thus say that an NMOS pass transistor passes a strong 0 and a weak 1. A similar argument will show that a PMOS pass transistor passes a weak 0 and a strong 1. When two NMOS pass transistors are chained in serial, a logic 1 will lose $2V_{tn}$ at the output. Thus, the length of a homogeneous pass transistor chain has to be limited for proper logic operation. The weak logic can be boosted by connecting the output to a CMOS inverter. Alternatively, we can use the complementary pass transistor logic style in which both N and P pass transistors are used forming a transmission gate. Figure 8.6(a) gives an implementation of a two-input multiplexor using complementary pass transistors.

Pass transistors are inherently suitable for implementing "multiplexor-like" Boolean functions. The pass transistor circuit in Figure 8.6(a) implements the Boolean function $Y = \overline{S}\,\overline{A} + S\overline{B}$. The same function implemented in static CMOS logic is shown in Figure 8.6(b). The pass transistor circuit requires 8 transistors while the static CMOS implementation requires 10 transistors. If we use the NMOS pass transistor implementation similar to that in Figure 8.5, only 6 transistors are required. In general, pass transistor logic is more efficient in implementing multiplexor-type and exclusive-OR type Boolean functions. It is from this observation that some designers believe that pass transistor logic could lead to a better power, delay and area implementation compared to the conventional static CMOS logic.

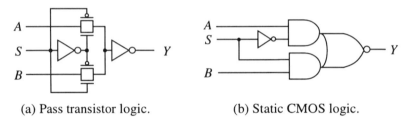

(a) Pass transistor logic. (b) Static CMOS logic.

FIGURE 8.6 Implementation of $Y = \overline{S}\,\overline{A} + S\overline{B}$.

8.2.2 Boolean Decision Diagram and Pass Transistor Logic

Before we discuss pass transistor logic synthesis system, we will introduce the *Boolean Decision Diagram* (BDD) as an alternate method to represent a Boolean function. A BDD consists of nodes labeled by the input variables of the Boolean function. An example is shown in Figure 8.7(c). To determine the output of the Boolean function, we traverse the BDD from the top down. When we reach a circular node, we ask whether the input variable of the node is at logic 0 or 1. We traverse the left edge downwards if the variable is logic 0; otherwise, we traverse the right edge. At the bottom of the diagram, we have two special square nodes representing logic 0 and 1, respectively. When the traversal reaches the special nodes, the logic value of the special node is the output of the function. For example in Figure 8.7(c), if both inputs of $Y = \overline{A} \cdot B$ are at logic 1, we follow the right links at both circle nodes and reach a logic 0.

A BDD is an alternate method to represent a Boolean function, just like a Boolean Equation or a Truth Table, as shown in Figure 8.7. There are several properties of BDD that are suitable for design automation applications but not found in some other representations. The storage requirement of a Truth Table is always exponential with

$$Y = \overline{A \cdot B}$$
$$Y = \overline{A} + \overline{B}$$

(a) Boolean equation.

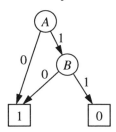

(c) Boolean Decision Diagram.

A	B	Y
0	0	1
0	1	1
1	0	1
1	1	0

(b) Truth table.

FIGURE 8.7 Different representations of the Boolean function $Y = \overline{A \cdot B}$.

respect to the number of inputs. This obviously presents a problem in handling the table in computer software. A Boolean Equation is not unique in the sense that many different equations can represent the same Boolean function. This also poses some problems in software because the outcome of the software execution on two equivalent Boolean Equations may not be the same when they should. It can be shown that BDD is unique when the ordering of the input variables (i.e., the ordering of the variables in the diagram) is given. This canonical property is desirable for software manipulation. Also for most Boolean functions, the size of the BDD is linear with respect to the number of input variables. Therefore, BDD is a very popular data structuring technique in design automation tools. In certain rare classes of Boolean functions, for example an exclusive-OR tree, the BDD can still have an exponential size.

The BDD of an arbitrary Boolean function can be readily constructed using recursive Shannon's decomposition, as shown in Figure 8.8. Note that each level of a BDD corresponds to the Shannon's expansion of an input variable. When a sub-function expanded is identical to an existing one, we simply link the tree branch to the existing subtree in the diagram rather than creating two identical subtrees. This results in the storage efficiency of BDDs. For example, if we expand the majority function $y = AB + BC + AC$ shown in Figure 8.10, the C node is shared by two edges from the higher level.

The very nature of a BDD lends itself to efficient multiplexor implementation, suitable for pass transistor logic. The downward traversal of the BDD at each node to determine its output can be mapped to a decision logic implemented by a two-input

$$f(x_1, \ldots, x_i, \ldots, x_n) \ = \ x_i f(x_1, \ldots, 1, \ldots, x_n) + \overline{x_i} f(x_1, \ldots, 0, \ldots, x_n)$$

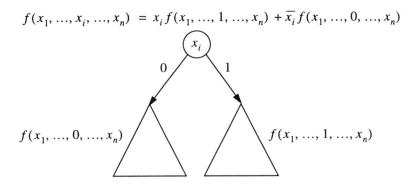

FIGURE 8.8 Using Shannon's decomposition to construct BDD.

multiplexor. The select signal of the multiplexor is connected to the input variable of the node. If the select signal is at logic 0 (logic 1), the output of the multiplexor is taken from the left (right) input because the left (right) subtree computes the output of the expanded Boolean sub-function. At the end of the expansion, the multiplexor input is either tied to logic 0 or 1. The mapping of BDD to multiplexor implementation is shown in Figure 8.9.

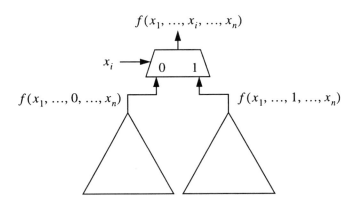

FIGURE 8.9 Implementing BDD using two-input multiplexors.

For example, the BDD and its two-input multiplexor mapping of the majority function $Y = AB + BC + AC$ are shown in Figure 8.10. The mapping of BDD nodes to multiplexors serves as a basis to generate logic circuits from arbitrary Boolean functions represented in BDDs. A direct mapping from a BDD to 2-input multiplexors may be area inefficient. Some heuristic manipulation of the BDD can be applied before the mapping to improve the area efficiency [8.8]. Since pass transistor logic is very efficient in implementing multiplexors, the logic circuits derived directly from this method are generally quite efficient. The pass transistor circuits generated are believed to have better power and area efficiency.

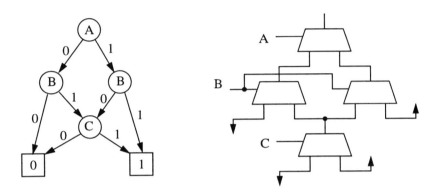

FIGURE 8.10 BDD of $Y = AB + BC + AC$ and its multiplexor implementation.

8.2.3 Pass Transistor Logic Synthesis System

The ability to compute arbitrary Boolean functions using direct mapping of BDDs to multiplexor-based pass transistor logic has important implications. Not only does it allow us to implement any combinational logic using pass transistors, it also opens the door for automated synthesis of pass transistor logic from hardware description language. Many logic synthesis systems in use today are designed for static CMOS type cells, which consist of NAND, NOR and CMOS complex gates as basic functional building blocks. Pass transistor logic is only scarcely used in some XOR or adder cells because of the particular Boolean functions encountered.

A pass transistor logic synthesis system has been reported in [8.8]. The basic operation of the synthesis system is to express Boolean logic in BDDs. The BDDs are then partitioned so that the length of the pass transistor chain is limited to avoid the threshold voltage drop problem mentioned in Section 8.2.1. The partitioned BDDs are then mapped into pass transistor cells and a post-mapping cleanup eliminates redundant circuits. The cell library consists of cells built with pass transistor circuits. An example of a library cell and its synthesis usage is shown in Figure 8.11. A unique property of most pass transistor cells is that different Boolean functions can be implemented by connecting the inputs differently. Figure 8.12 shows two different Boolean functions implemented by the same pass transistor cell.

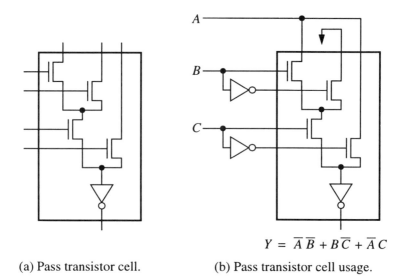

$$Y = \overline{A}\,\overline{B} + B\,\overline{C} + \overline{A}\,C$$

(a) Pass transistor cell.   (b) Pass transistor cell usage.

FIGURE 8.11 A Pass transistor cell in logic synthesis.

Some experimental results published in [8.8] have shown that using the proper logic synthesis algorithms, pass transistor logic achieves better power and area efficiency than the static CMOS logic. Many aspects of the synthesis algorithms still need to be improved. The incorporation of mixed static CMOS and pass transistor logic cell libraries need to be investigated and the threshold voltage drop problem in extremely low voltage circuits has not been well studied. The wide spread acceptance of synthesized pass transistor logic in the mainstream digital logic design remains to be seen.

8.3 Asynchronous Circuits

The mainstream VLSI design today are mostly *synchronous* systems. The operation of a synchronous system is governed by a *clock* signal that typically controls all sequential cells. The clock signal carries no information nor does it participate in logic computation. It only provides synchronization to the system so that the compu-

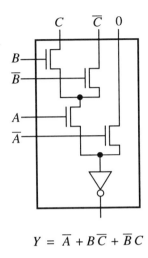

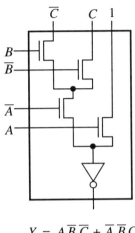

$$Y = \overline{A} + B\overline{C} + \overline{B}\,C \qquad\qquad Y = A\overline{B}\,\overline{C} + \overline{A}\,\overline{B}\,C$$

FIGURE 8.12 Pass transistor cell implementing different Boolean functions.

tation can be performed in a lock-step fashion. The clock signal needs to drive thousands of sequential cells and it is also the highest frequency signal in the system. Therefore, its power dissipation represents a large portion of the total system power consumption as already mentioned in Section 6.1.

The *asynchronous* system [8.9] is a type of computation system without any global clock signal. Computation is achieved by a series of *events* represented by signal transition edges. The proper *sequence* of signal edges is required to achieve the desired computation. However, the duration it takes from one signal edge to another does not affect the correctness of the computation. Therefore, the speed of computation is determined by the signal propagation delay of the asynchronous circuit, not a global clock signal.

8.3.1 Asynchronous System Basics

Asynchronous circuits have been widely used in bus handshaking protocols. A simple example of an asynchronous system is the CPU interrupt request and acknowledge mechanism, illustrated in Figure 8.13. When a peripheral device needs to notify the CPU of a condition, it sends out the appropriate interrupt vector (which contains information about the interrupt condition) and asserts the interrupt request signal to low. The peripheral device does not share a common clock with the CPU nor does it know *when* the interrupt request will be served. When the CPU is ready to handle the interrupt request, it pulls the interrupt acknowledge signal to low, indicating that it is reading the interrupt vector. After the completion of the information transfer, the CPU restores the interrupt acknowledge signal. Upon detecting the rising edge of the interrupt acknowledge signal, the peripheral device realizes that the CPU has gotten the vector information; removes the interrupt request and gets ready for the next interrupt.

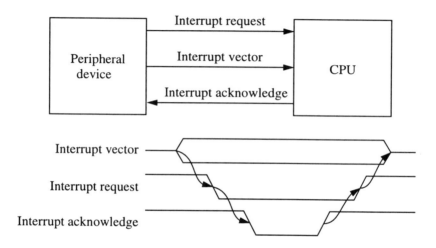

FIGURE 8.13 Interrupt request and acknowledge protocol.

The following sequence of events occurs asynchronously:

1. Peripheral device sends out the interrupt vector and interrupt request.
2. CPU acknowledges the request, reads the interrupt vector and reacts to the interrupt condition.
3. CPU signals the completion of interrupt handling.
4. Peripheral device removes the interrupt request and vector.

The arrows in the timing diagram show the cause and effect relationship of the sequence of events. Note that there is no clock signal required to coordinate the timing of the events. However, both parties have to act and react according to a predetermined protocol to achieve the desired computation, which, in this case, is a simple vector transfer.

The same scheme can be used between modules in the datapath of a complex computation system. For example, we can replace the peripheral device and the CPU with a multiplier and an adder, respectively. The output of the multiplier is fed to the adder and as soon as the adder completes its computation, it signals the multiplier to remove the data and send the next wave of data.

One architecture of an asynchronous processing unit is shown in Figure 8.14. The delay line is set to the worst case signal delay of the combinational logic unit. After the input data have been processed by the combinational logic unit, the controller latches the computation results; asserts the request for the next stage and waits until it receives the acknowledge signal from the down stream unit before freeing the current output data.

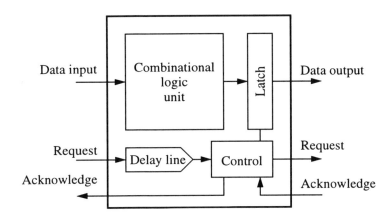

FIGURE 8.14 An asynchronous processing unit.

8.3.2 Prospects of Asynchronous Computation

The most obvious power saving from asynchronous computation is the elimination of the global clock signal. This avoids many problems associated with distribution of high frequency clock signals on the chip. The speed of an asynchronous system is only limited by the natural delay of the circuit elements and stringent timing synchronization is not required. So, an asynchronous system is more robust under delay variation. In some ways, the request and acknowledge signals act like the clock signal because they synchronize the order of computation. However, unlike a global clock signal, the switching of the request and acknowledge signals indicate the computation conditions of the local processing unit. These signals need not be routed throughout the chip, only between processing units, and there are no useless switching activities in the handshaking signals. Therefore, the power dissipation of an asynchronous system is believed to be better than the synchronous counterpart.

Since an asynchronous system is delay insensitive, it can be operated at very low voltage. The circuit and logic design of the system typically does not require detailed path delay analysis across different computation modules. The operating voltage can be fine-tuned to match the throughput requirement of the system. Recall Section 7.1.2, the operating voltage of the asynchronous processing unit is adjusted based on the workload of the system.

For control dominated systems such as a CPU, asynchronous implementation may not be desirable because of the complex handshaking between processing units. Data intensive computation such as DSP algorithms can be implemented with asynchronous logic. Like pass transistor logic, automatic logic synthesis of asynchronous systems is not widely available. This somewhat hampers the design productivity. The testing issues of asynchronous systems are not well studied and the area overhead of the control and handshaking logic could be too high.

The implementation of an asynchronous chip for error correction code application has been demonstrated [8.10]. It reportedly achieves an impressive 80% power reduction compared to the synchronous implementation. The area overhead was estimated to be 70-100% more, mainly due to the use of two signal lines to represent a bit.

REFERENCES

[8.1]   W. Athas, L. Svensson, J. Koller, N. Tzartzanis and E. Chou, "Low-Power Digital Systems Based on Adiabatic-Switching Principles," *IEEE Transactions on VLSI Systems*, vol. 2, no. 4, pp. 398-407, Dec. 1994.

[8.2]   J. Denker, S. Avery, A. Dickinson, A. Kramer and T. Wik, "Adiabatic Computing with the 2N-2N2D Logic Family," *Proceedings of International Workshop on Low Power Design*, pp. 183-187, 1994.

[8.3]   A. Kramer, J. Denker, B. Flower and J. Moroney, "Second Order Adiabatic Computation with 2N-2P and 2N-2N2P Logic Circuits," *Proceedings of International Symposium on Low Power Design*, pp. 191-196, 1995.

[8.4]   T. Indermaur and M. Horowitz, "Evaluation of Charge Recovery Circuits and Adiabatic Switching for Low Power CMOS Design," *Digest of Technical Papers, IEEE Symposium on Low Power Electronics*, pp. 102-103, 1994.

[8.5]   D. Frank, "Comparison of High Speed Voltage-Scaled Conventional and Adiabatic Circuits," *Digest of Technical Papers, International Symposium on Low Power Electronics and Design*, pp. 377-380, 1996.

[8.6]   M. Knapp, P. Kindlmann and M. Papaefthymiou, "Implementing and Evaluating Adiabatic Arithmetic Units," *Proceedings of Custom Integrated Circuit Conference*, pp. 115-118, 1996.

[8.7]   D. Somasekhar, Y. Ye and K. Roy, "An Energy Recovery Static RAM Memory Core," *Digest of Technical Papers, IEEE Symposium on Low Power Electronics*, pp. 62-63, 1995.

[8.8]   K. Yano, *et al.*, "Top-Down Pass-Transistor Logic Design," *IEEE Journal of Solid-State Circuits*, vol. 31, no. 6, pp. 792-803, 1996.

[8.9]   T. Meng, *Asynchronous Circuit Design for VLSI Signal Processing*, Kluwer Academic Publishers, Boston, 1994.

[8.10] K. Berkel, *et al.*, "A Fully Asynchronous Low Power Error Corrector for the DCC Player," *Digest of Technical Papers, International Solid-State Circuit Conference*, pp. 88-89, 1994.

Index